S. B Higgins

Ophidians, Zoological Arrangement of the Different Genera

Including Varieties Known in North and South America

S. B Higgins

Ophidians, Zoological Arrangement of the Different Genera
Including Varieties Known in North and South America

ISBN/EAN: 9783337123680

Printed in Europe, USA, Canada, Australia, Japan

Cover: Foto ©berggeist007 / pixelio.de

More available books at **www.hansebooks.com**

OPHIDIANS,

ZOOLOGICAL ARRANGEMENT OF THE DIFFERENT GENERA,

INCLUDING

VARIETIES KNOWN IN NORTH AND SOUTH AMERICA, THE EAST INDIES, SOUTH AFRICA, AND AUSTRALIA.

THEIR POISONS,

AND ALL THAT IS KNOWN OF THEIR NATURE.

THEIR GALLS,

AS ANTIDOTES TO THE SNAKE-VENOM.

PATHOLOGICAL, TOXICOLOGICAL, AND MICROSCOPICAL FACTS;
TOGETHER WITH MUCH INTERESTING MATTER HITHERTO NOT PUBLISHED.

BY

S. B. HIGGINS, S. A.,

HONORARY MEMBER OF THE HOMŒOPATHIC INSTITUTE OF THE UNITED STATES OF COLOMBIA.

FIRST AMERICAN EDITION.

PUBLISHED BY
BOERICKE & TAFEL,

NEW YORK: PHILADELPHIA:
No. 145 GRAND STREET.   No. 635 ARCH STREET.

1873.

DEDICATORY.

TO DR. ANTONIO M. BUITRAGO,

OF BARANQUILLA, UNITED STATES OF COLOMBIA,

AS A TESTIMONIAL OF ESTEEM AND FRIENDSHIP,

This Volume is Dedicated,

IN APPRECIATION OF THOSE STERLING QUALITIES WHICH

COMBINE TO MAKE A TRUE PHYSICIAN,

AND WHICH HE POSSESSES IN AN EMINENT DEGREE:

BY THE AUTHOR.

INTRODUCTION.

The subject treated in this work has occupied the author's attention and study for many years. When quite young he witnessed the excruciating torments suffered by a person bitten by a rattlesnake, and the question suggested itself to his mind, Can it be possible that such a deadly poison exists in nature without an antidote existing *somewhere near it* at the same time?

In Bible history,* when the Israelites, journeying from Mount Hor by the way of the Red Sea, were visited by the plague of "numberless fiery serpents," Moses, at God's command, caused to be made "a serpent of brass, and put it upon a pole, and it came to pass that if a serpent had bitten any man, when he beheld the serpent of brass he lived." This seems to indicate that the antidote to the poison lies in the snake itself. Following up this idea leads us to the bile or gall as the object of our search.

Experiments prove that its efficacy as an antidote is beyond dispute. A very great number of cures of snake-bites made by preparations of snake-galls place a seal upon the value of this discovery.

The Author makes no claim to literary merit in this work, but it is given to the public with the hope that the student may find

---

* Numbers, Chapter XXI.

in its pages more matter than has ever yet been collated into one volume upon the subject, Ophidians; that the physician may find herein in a single mass of facts all that is as yet known, and all the reliable matter hitherto published in any work about snake-poisons and their action in a pathological and in a toxicological sense; and some facts about their nature which have never been mentioned heretofore by any one who has made the subject a study; and that the general reader may find, in the ensuing pages, wherewith to pass an agreeable hour, and herein to add some new and interesting facts to his store of general information.

As yet but little is known about the action of serpent-poisons, so that every contribution to science in this sense adds to the knowledge already acquired. The Author hopes that if anything really new is contributed in this volume it may serve as an incentive to some one else, better qualified by scientific attainments than himself, to make further and more exhaustive studies in this branch; and that this interest may not cease until our knowledge on this point shall have become complete in every respect.

In cold climates, where few really venomous snakes exist, this subject is of little general interest; but in the tropics, where one is everywhere, both by night and by day, liable to be bitten by one of these repugnant reptiles, it becomes a matter of life and death.

In pursuing this study I had occasion to visit London, and while there had access to the British Museum Library. This is an institution of which England may boast with pride.

I went there a perfect stranger, and after going through a simple formality I was surrounded with every possible convenience and attention by the admirable regulations, so well and efficiently carried out by the gentlemanly superintendent of the reading-room, Mr. Bull, and a host of attendants under him. Any one of the *million* volumes stored upon the shelves was at my disposition, and after pondering over some of them to my liking, I came away with one poignant regret, viz., that in America we have nothing that even approaches to this vast treasure-store of the world's literature; that on this point England stands alone, unapproachable!

I desire, in this connection, to express my obligations to Dr. J. Forbes Watson, of the India Office, for his obliging and polite attentions in many ways: to Lord Calthorpe, for using his influence in my favor with the Board of the Zoological Society of London: to Lord Odo Russell, British Ambassador to Berlin, for letters of introduction so kindly furnished by him: and to Robert Bunch, Esq., Her Majesty's Ambassador to Bogota, United States of Colombia, for letters to Earl Granville, &c., &c.

I permit myself to indulge the hope that, although this task has been a repugnant one, and attended by such great risks as no one knows of who has not wrought in the same field, yet directions herein given may enable some one who peruses this work to save thereby some fellow-mortal's life. To know of such a case will more than repay my efforts, and afford me the consolation which follows after the performance of every labor of love.

S. B. HIGGINS.

GREENSBORO, NORTH CAROLINA,
    January, 1873.

PREFATORY.

The Author does not claim to have been the first person to use galls of a venomous serpent to cure bites of the same, for it is known that the Curers of greatest fame in Venezuela have used a mixture of galls with other antidotes for many years; and it is believed that the use of this mixture was practiced by the Indians, and by them communicated to the former, possibly so long as a century ago.

He does claim, however, to have been the first person to use the gall alone, unmixed with any other substance, to cure the bite of a serpent of the same kind from which the gall was taken; and also to have been the first person to initiate and carry out a long series of experiments with snake-poisons, the results of which have developed to him a new law in therapeutics, which may be expressed in the following terms: "*Every animal poison has its perfect and specific antidote in the gall of the animal or reptile in which that poison is secreted.*"

The study of this subject was commenced in the year 1850, and when he had an opportunity to make the experiments just referred to, his only object being to discover, if possible, a specific antidote for snake-poisons, without a thought of putting his labors in a book form; no records of the experiments were kept, and no data retained, except in memory, which can be given in proof of what is herein asserted.

After having cured nearly fifty persons bitten, and at the instance of several friends in South America, he was persuaded to publish a small work in Spanish, intended to give some general notions about Ophidians, a list of antidotes, and the "secret methods of cure" used by the Curers, &c., &c., intended as a manual, and containing the method of preparation of the galls, and such simple directions about their administration as could be comprehended clearly by every person of the lowest mental capacity, and which would enable such a person to treat any case of snake-bite without hesitation. A continued and further study of the subject, and a desire to extend his field of labors in this sense, led to his placing his discoveries at the disposition of the British Government.

Further studies and researches made at the great library of the British Museum, that greatest of all the storehouses of scientific lore in the world, have given the subject the form in which it is here presented to the public. No one is more fully sensible than himself of its imperfections. His desire was to have made the study exhaustive, and to this end further studies will be prosecuted, which he hopes at some future time may be received by those who take any interest in the subject with the same leniency he asks for this humble literary effort, made with the hope that it may not be devoid of value to science.

OPHIDIANS.

GENERAL REMARKS.

In ancient times serpents were an object of worship. Las Casas speaks of the Ringed Boa, or Aboma, found in Mexico, a reptile of great destructive powers, as being worshipped by them as an object of fear. The Incas considered the snake to be a symbol of cunning and wisdom, and as such it was sacred and adored. The Hindoos believed it to be possessed of supernatural powers, of which they saw *glimpses in the expanded hood of the Cobra*, consequently images of Brahma and Vishnu are found in their temples and on the decorations of the cars of Juggernaut, standing upon the caudal coil of a serpent whose perpendicular folds overlie each other, and whose neck and head expand into a broad concavo-convex hood, whose upper periphery is divided into triangular points, each of which bears a smaller head with distended jaws, thus forming the appearance of a canopy over the head of the god. A small image carved in stone, and about three feet in height, answering nearly to the above description, may be seen at the India Museum, in the India Office, in London.

The symbolical twining of serpents around the staff of Æsculapius indicates a belief in its endowment with some extraordinary virtues in connection with the healing art, while the Caduceus of Mercury, composed of two intertwined serpents, can be supposed to convey a subtler meaning, and symbolize a belief in its magnetic power; its bearer was the messenger of the gods, and his wand may be supposed to typify an attribute of a higher nature not known to its bearers, but believed to exist in the reptile.

To go back to the remotest epoch in history, we find the prince of darkness represented as a serpent tempting Eve; and it should cause no wonder to the student, who tries to trace popular beliefs to their origin, that all the histories of all people and nations should bear some record of the adoption of a marked belief in the existence of some peculiar virtues or characteristics in these reptiles. These are born of experience, and merit of the man of science, a profound attention and study, although the scientific world has heretofore satisfied itself by assigning to such beliefs a purely superstitious origin.

The classification of Ophidians is a task attended with great difficulty in many cases.

Some varieties are distinctly marked, such as the Elaps Corallinus, for example; others occur possessing very different characteristics; a third class possess marks peculiar to both species, produced by a male of the first kind copulating with a female of the latter species: this leads to the formation of a new species, and following this custom of naturalists leads to a most interminable number of varieties, which could be very much simplified by adopting another method of classification.

BUFFON'S CLASSIFICATION.

Buffon's classification of Ophidians* is as follows, viz.:

Genera.	No. of species.	Genera.	No. of species.	Genera.	No. of species.
1. Boa,	18	10. Cenchris,*	1	19. { Orvetia, Anguis, }	16
2. Python,	5	11. Vipera,*	5	20. Ophiosaurus,	2
3. Coral,	1	12. Coluber,	150	21. Pelamis,	3
4. Bungarus,*	2	13. Platurus,*	4	22. Hydrophis.*	6
5. Hurria,	3	14. Enhydris,	5	23. Acrochordon,	1
6. Acanthopis,*	1	15. Langaha,*	1	24. Amphisbœna,	2
7. Crotalus,*	7	16. Erpetons,	1	25. Cecilia,	4
8. Scytalus,*	5	17. Eryxæ,	4		
9. Lachesis,*	2	18. Clothonia,*	2		

Making a total of 25 genera composed of 251 species; of these he considered 11 genera (viz., Nos. 4, 6, 7, 8, 9, 10, 11, 13, 15, 18, and 22), composing 36 species, as venomous, and 14 genera of 215 species as innocuous.

Many other species of the genera innocuous are known to be venomous, such as the Coral, Acrochordon, and Amphisbœna, and later naturalists have found it necessary to modify the preceding classification considerably. The one generally accepted as the most complete at the present time is undoubtedly that of the celebrated zoologist, Dr. Günther.†

Dr. Fayrer, in his Thanatophidia, adopts Günther's classification. This gives no North American genera. The discrepancy, however, is filled by the following:

---

* Buffon's Nat. Hist., vol. 83, p. 87.
† Reptiles of British India, London, 1864.

SYSTEMATIC TABLE OF WELL-ASCERTAINED SPECIES OF NORTH AMERICAN SERPENTS.*

First Genus—CROTALUS, *Linn.*

1. Crotalus durissus, *Linn.*   Penna., Louisiana, Miss.
2. "   adamantus, *Beauv.*   South Carolina.
3. "   atrox, *B. & G.*   Texas.
4. "   lucifer, *B. & G.*   Oregon.
5. "   confluentus, *Say.*   Arkansas, Texas.
6. "   molossus, *B. & G.*   Sonora.
7. "   Oregonus, *Holbr.*   Columbia River.

Second Genus—CROTALOPHORUS, *Gray.*

1. Crotalophorus miliarius, *Holbr.*   Georgia, S. Carolina.
2. "   consors, *B. & G.*   Texas.
3. "   tergeminus, *Holbr.*   Wis., Mich., Ohio.
4. "   Edwardsii, *B. & G.*   Mexico, Sonora.
5. "   Kirtlandii, *Holbr.*   Ohio.

Third Genus—AGKISTRODON, *Beauv.*

1. Agkistrodon contortrix, *B. & G.*   Ohio, Pa., S. C., La.

Fourth Genus—TOXICOPHIS, *Troost.*

1. Toxicophis piscivorus, *B. & G.*   Louisiana.
2. "   pugnax, *B. & G.*   Texas.

Fifth Genus—ELAPS. *Fitz.*

1. Elaps fulvius, *Cuv.*   South Carolina.
2. "   tenere, *B. & G.*   Texas.
3. "   tristis, *B. & G.*   Mississippi, Texas.

---

* Catalogue of North American Reptiles, by Baird and Girard. Publications by the Smithsonian Institute, Washington, 1853.

Sixth Genus—EULAINIA, *B. & G.*

1. Eulainia saurita, *B. & G.*   Mass., Pa., N. Y., Md., Va.
2. " Faireyi, *B. & G.*   Louisiana.
3. " proxima, *B. & G.*   Ark., Texas, New Mexico.
4. " infernalis, *B. & G.*   California.
5. " Pickeringii, *B. & G.*   Oregon.
6. " parietalis, *B. & G.*   Texas.
7. " leptocephala, *B. & G.*   Oregon.
8. " sirtalis, *B. & G.*   Me., Mich., N. Y., Penna., Md., Va., S. C., Miss.
9. " dorsalis, *B. & G.*   Texas.
10. " ordinata, *B. & G.*   Georgia.
11. " ordinoides, *B. & G.*   California.
12. " radix, *B. & G.*   Wisconsin.
13. " elegans, *B. & G.*   California.
14. " vagrans, *B. & G.*   Mexico, California, Oregon.
15. " Marciana, *B. & G.*   Arkansas, Texas.
16. " concinna, *B. & G.*   Oregon.

Seventh Genus—NERODIA, *B. & G.*

1. Nerodia sipedon, *B. & G.*   Mich., Mass., Pa., N. Y., Md.
2. " fasciata, *B. & G.*   South Carolina.
3. " erythrogaster, *B. & G.*   Louisiana, S. Carolina.
4. " Agassizii, *B. & G.*   Lake Huron.
5. " Woodhousii, *B. & G.*   Texas.
6. " taxispilota, *B. & G.*   Georgia.
7. " Holbrookii, *B. & G.*   Louisiana.
8. " niger, *B. & G.*   Massachusetts.
9. " rhombifer, *B. & G.*   Arkansas.
10. " transversa, *B. & G.*   Arkansas.

Eighth Genus—REGINA, *B. & G.*

1. Regina leberis, *B. & G.*   Mich., Ohio, Pennsylvania.
2.   "   rigida, *B. & G.*   Pennsylvania, Georgia.
3.   "   Grahamii, *B. & G.*   Texas.
4.   "   Clarkii, *B. & G.*   Texas.

Ninth Genus—NINIA, *B. & G.*

1. Ninia diademata, *B. & G.*   Mexico.

Tenth Genus—HETERODON, *Beauv.*

1. Heterodon platyrhinos, *Latr.*   Pa., Va., S. C., Ohio, Miss.
2.   "   cognatus, *B. & G.*   Texas.
3.   "   niger, *Troost.*   Pa., S. Carolina, Mississippi.
4.   "   atmodes, *B. & G.*   Georgia, S. Carolina.
5.   "   simus, *Holbr.*   S. Carolina, Mississippi.
6.   "   nascicus, *B. & G.*   Ark., Texas, Sonora, Cal.

Eleventh Genus—PITUOPHIS, *Holbr.*

1. Pituophis melanoleucus, *Holbr.*   California.
2.   "   bellona, *B. & G.*   Texas, California, Sonora.
3.   "   McClellanii, *B. & G.*   Arkansas.
4.   "   catenifer, *B. & G.*   California.
5.   "   Wilkesii, *B. & G.*   Oregon.
6.   "   annectens, *B. & G.*   California.

Twelfth Genus—SCOTOPHIS, *B. & G.*

1. Scotophis Alleghaniensis, *B. & G.*   Pennsylvania.
2.   "   Lindheimerii, *B. & G.*   Texas.
3.   "   vulpinus, *B. & G.*   Michigan, Wisconsin.
4.   "   confinis, *B. & G.*   South Carolina.

SPECIES OF NORTH AMERICAN SERPENTS. 17

5. Scotophis lætus, *B. & G.* Arkansas.
6. " guttatus, *B. & G.* S. C., Ga., Miss.
7. " quadrivittatus, *B. & G.* Florida.
8. " Emoryi, *B. & G.* Texas.

Thirteenth Genus—OPHIBOLUS, *B. & G.*

1. Ophibolus Boylii, *B. & G.* California.
2. " splendidus, *B. & G.* Sonora.
3. " Sayi, *B. & G.* La., Miss., Ark., Texas.
4. " getulus, *B. & G.* S. C., Miss.
5. " rhombomaculatus. *B. & G.* Ga., S. C.
6. " eximius, *B. & G.* Mass., N. Y., Penn.
7. " clericus, *B. & G.* Va., Miss.
8. " doliatus, *B. & G.* Mississippi.
9. " gentilis, *B. & G.* Arkansas, Louisiana.

Fourteenth Genus—GEORGIA, *B. & G.*

1. Georgia Couperi, *B. & G.* Georgia.
2. " obsoleta, *B. & G.* Texas.

Fifteenth Genus—BASCANION, *B. & G.*

1. Bascanion constrictor, *B. & G.* Pa., Md., Miss., S. C., La.
2. " Fremontii, *B. & G.* California.
3. " Foxii, *B. & G.* Michigan, Pennsylvania.
4. " flaviventris, *B. & G.* Texas, California.
5. " vetustus, *B. & G.* California, Oregon.

Sixteenth Genus—MASTICOPHIS, *B. & G.*

1. Masticophis flagelliformis, *B. & G.* South Carolina.
2. " flavigularis, *B. & G.* Texas, Arkansas.
3. " mormon, *B. & G.* Utah.

4. Masticophis ornatus, *B. & G.* Texas.
5.   "   taeniatus, *B. & G.* California.
6.   "   Schottii, *B. & G.* Texas.

Seventeenth Genus—SALVADORA, *B. & G.*

1. Salvadora Grahamiæ, *B. & G.* Sonora.

Eighteenth Genus—LEPTOPHIS, *Bell.*

1. Leptophis æstivus, *Bell.* Md., Va., S. C., Miss.
2.   "   majalis, *B. & G.* Texas, Arkansas.

Nineteenth Genus—CHLOROSOMA, *Wagl.*

1. Chlorosoma vernalis, *B. & G.* Maine, Mass., N. Y., Penn., Mich., Wisconsin, Miss.

Twentieth Genus—CONTIA, *B. & G.*

1. Contia mitis, *B. & G.* California, Oregon.

Twenty-first Genus—DIADOPHIS, *B. & G.*

1. Diadophis punctatus, *B. & G.* N. Y., Pa., Ga., S. C., Mississippi.
2.   "   amabilis, *B. & G.* California.
3.   "   docilis, *B. & G.* Texas.
4.   "   pulchellus, *B. & G.* California.
5.   "   regalis, *B. & G.* Sonora.

Twenty-second Genus—LODIA, *B. & G.*

1. Lodia tenuis, *B. & G.* Oregon.

Twenty-third Genus—SONORA, *B. & G.*

1. Sonora semiannulata, *B. & G.* Sonora.

Twenty-fourth Genus—RHINOSTOMA, *Fitz.*
1. Rhinostoma coccinea, *Holbr.* S. C., Ga., Miss., La.

Twenty-fifth Genus—RHINOCHEILUS, *B. & G.*
1. Rhinocheilus Lecontii, *B. & G.* California.

Twenty-sixth Genus—HALDEA, *B. & G.*
1. Haldea striatula, *B. & G.* Va., S. C., Miss.

Twenty-seventh Genus—FARANCIA, *Gray.*
1. Farancia abacurus, *B. & G.* South Carolina, Louisiana.

Twenty-eighth Genus—ABASTOR, *Gray.*
1. Abastor erythrogrammus, *Gray.* Georgia.

Twenty-ninth Genus—VIRGINIA, *B. & G.*
1. Virginia Valeriæ, *B. & G.* Md., Va., S. C.

Thirtieth Genus—CELUTA, *B. & G.*
1. Celuta amœna, *B. & G.* Pa., Md., Va., S. C., Miss.

Thirty-first Genus—TANTILLA, *B. & G.*
1. Tantilla coronata, *B. & G.* Mississippi.
2. " gracilis, *B. & G.* Texas.

Thirty-second Genus—OSCEOLA, *B. & G.*
1. Osceola elapsoidea, *B. & G.* S. C., Miss.

Thirty-third Genus—STORERIA, B. & G.

1. Storeria Dekayi, *B. & G.* Wis., Mich., Ohio, Mass., N. Y., Pa., Md., S. C., Ga., La., Texas.
2. " occipito-maculata, *B. & G.* Maine, N. Y., Lake Superior, Wis., Pa., S. C., Ga.

Thirty-fourth Genus—WENONA, B. & G.

1. Wenona plumbea, *B. & G.* ·Oregon.
2. " Isabella, *B. & G.* Oregon.

Thirty-fifth Genus—RENA, B. & G.

1. Rena dulcis, *B. & G.* Texas.
2. " humilis, *B. & G.* California.

SUMMARY.

Genera,	35
From below,	2
Total genera,	— 37
Species,	119
From below,	9
Total species,	—128

OTHER SPECIES, NOT INCLUDED IN THE PRECEDING.

1. Toxicophis atrofuscus, *Troost.* Tennessee.
2. Coluber testaceus, *Say.* Rocky Mountains.
3. " Sayi, *Schlegel.* Missouri.
4. " vertebralis, *Blainv.* California.
5. " (Ophis) Californiæ, *Blainv.* California.
6. " (Zacholus) zonatus, *Blainv.* California.
7. " planiceps, *Blainv.* California.

8. Coluber Charina Bottæ, *Gray*. (Boidæ.) California.
9. Ophthalmidion longissimum, *Dum. & Bibr.* (Typhlopidæ.) Florida.

These 37 genera, composed of 128 species, are embraced under 4 family groups, with the following specific and distinguishing characteristics, viz.:

FAMILY I.—CROTALIDÆ. Erectible poison-fangs in front. Few teeth in upper jaw. A deep pit between eye and nostril.

FAMILY II.—COLUBRIDÆ. Both jaws fully provided with teeth. No anal appendages.

   A. Loreal and anteorbital shields both present.
   B. Either loreals or anteorbitals absent.

FAMILY III.—BOIDÆ. Both jaws with teeth. Rudiments of hinder limbs, or spur-like anal appendages.

FAMILY IV.—TYPHLOPIDÆ. Teeth only in one jaw, either the upper or lower. Upper jaw prominently projecting. Scales on the belly instead of scutellæ scuta?, disposed in several series, like those on the dorsum.

FAMILY I comprises Genera Nos. 1 to 4, inclusive.
    II comprises Genera Nos. 5 to 33, inclusive.
    III comprises Genera Nos. 34 and 35.
    IV comprises Genera Nos. 36 and 37.

Few of the preceding species are included in Günther's classification, but more are referred to in descriptions of species in this work.

The classification just mentioned embraces only 21 genera, which are composed of 109 species, and is as follows, viz.:

No.	Genera.	No. of Species.	No.	Genera.	No. of Species.
1.	Typhlopidæ,	8	12.	Dipsadidæ,	10
2.	Tortricidæ,	2	13.	Lycodontidæ,	11
3.	Zenopeltidæ,	1	14.	Amblycephalidæ,	4
4.	Uropeltidæ,	16	15.	Pythonidæ,	2
5.	Calamaridæ,	6	16.	Erycidæ,	3
6.	Oligodontidæ,	2	17.	Acrochordidæ,	3
7.	Colubridæ,	4	18.	Elapidæ.	2
8.	Homalopsidæ,	8	19.	Hydrophidæ,	7
9.	Psammophidæ,	2	20.	Crotalidæ,	4
10	Dendrophidæ,	5	21.	Viperidæ,	25
11.	Dryiophidæ.	6		Total No. of Species,	109

The classification by Baird and Girard of existing species found in North America is differently arranged from Günther's, and has such a complicated nature that it has few genera and fewer species in common with it.

As far, however, as it is possible to harmonize all the classifications of different naturalists herein given, a summary gives the following 8 families, viz.:

FAMILY I. COLUBRIDÆ.　　FAMILY V. ELAPIDÆ.
　　II. BOIDÆ.　　　　　　　　　VI. HYDROPHIDÆ.
　　III. TYPHLOPIDÆ.　　　　　VII. VIPERIDÆ.
　　IV. COLUBRIFORMES.　　VIII. CROTALIDÆ.

Composed of 227 genera and 645 species; of all of which latter mention is made in this work.

OPHIDIA COLUBRIFORMES.

Innocuous Snakes, without grooved or perforated fanglike teeth in front.

FIRST GENUS—TYPHLOPIDÆ.

Body cylindrical, rigid, covered with comparatively large, polished scales, firmly adherent; head not distinct from neck; none of the teeth are enlarged; ventral scales not larger than dorsal; no mental groove; upper labials, four.

Of this genus India and the Eastern world furnish 8 species, viz.:

    No. 1. Typhlina lineata. *Wagler.*
        2. Typhlops nigro-albus. *D. n. & B.*
        3.   "   Horsfieldii. *Gray.*
        4.   "   bothryorhynchus. *Günther.*
        5.   "   striolatus. *Peters.*
        6.   "   Siamensis. *Günther.*
        7.   "   Braminus. *Dum. Cuvier.*

This latter species is **the Rondoo Talooloo Pam** of Dr. P. Russell, and is described as follows, **under two varieties,** viz.:

Called R. T. Pam, or "double-headed snake." Abdominal squamæ, 251; subcaudal 120 = 271. Head not broader than the neck, ovate, obtuse, **convex,** covered with nine laminæ of unusual shapes, mouth contracted, marginal maxillary, and pterygoid rows of teeth in upper jaw; eyes lateral, small, not prominent; nostrils close, and very small; trunk cylindrical and of nearly equal thickness from head to tail; scales small, orbicular, imbricate, each with a black dot upon its apex; length 10½ inches, diameter less than ⅓ inch; tail 4½ inches; has a blunt point and very little taper; color reddish-brown, part of the tail is cineritious or very pale blue; the squamæ are glossy white.

OBSERVATIONS.

This variety is very **quick in** its movements, and, if loosened on the sand, it burrows into it instantly, and is probably called **two-headed** from the fact that it moves forward and backward **with** equal facility.*

Another variety, called R. T. Pam, in Vizagapatam is 6 inches in length; head only distinguishable upon close inspection, as the body is

---

    * For this reason it should be classed among the Amphisbæna.

of a uniform thickness, and point of tail very obtuse. There are 3 or 4 very small laminæ on the head, which is covered with suborbicular scales, too small to be counted; *diameter* hardly ¼ inch; *color*, cream, with innumerable black dots.*

OBSERVATIONS.

Very common in Vizagapatam, and crawls backward and forward with equal facility.

No. 8. Typhlops onychocephalus. *Wagl.* Australia furnishes 10 species, viz.:

Typhlops polygramaticus. Schlegel's blind snake. *Krefft.*†
" nigrescens. Gray's blind snake.
" Rüppellii. Ruppell's blind snake.
" Preisii. Preiss's blind snake.
" bicolor. Schmidt's blind snake.
" bituberculatus. Peters's blind snake.
" Australis. West Australia blind snake.
" Güntherii. Günther's blind snake.
" Wiedii. Wied's blind snake.
" unguirostris. Queensland blind snake.

No species are classified under African genera.

In South America several species are known to exist, particularly in Brazil and in the forests of the Isthmus of Panama; some are also found in Venezuela.

Captain Raverty says,‡ in Pushto there is a species of snake called *Kaochah* or *Kawchah*, of a dirty earth color, covered with red spots, whose bite is said to be mortal. Individuals of this variety are said to be as thick as a man's arm, less than two feet in length, and of a very repulsive appearance.

---

* This is another Amphisbœnus.
† G. Krefft's Snakes of Australia.
‡ Notes on Kaferistan.

The *Gunnus* or *Aphia* is also very poisonous, of about the same size as the preceding, and of very repulsive appearance. Both these species are found in stony places: the latter particularly abounding along the eastern slope of the valley of Kashmir.

Wise says* that the Grecian physicians of olden times found no remedies for snake-bites, but the Indians did cure persons bitten; on which account an edict was published, ordering all persons bitten by a serpent to be brought to the king's tent, where the Hindoo physicians would cure them.

SECOND GENUS—TORTRICIDÆ.

Ventral scales but little larger than the others; a mental groove exists; upper labials six.

These are "short tails," those found in the East Indies belonging to one species only.

No. 1. *Tort. cylindrophis* (Wagler. Rufus. Gray), is the Schilay Pamboo of Dr. Patrick Russell.

No. 2. *T. cylindrophis maculatus* (Linn.) is the Anguis maculata of Russell. These are both East India species.

Wood† classifies the Coral snake as Tortrix scytale, but Wagler and other naturalists place it under the genus Elaps. A classification of the South American genera will find other varieties belonging to this genus.

No. 3. *Tortrix bottæ*—Charinæ bottæ—(Gray. Blaine.), is a North American variety.

THIRD GENUS—XENOPELTIDÆ Gray.

Ventrals distinct; two pair of frontals; five occipitals

Cantor gives one East India variety of this genus, viz.,

---

* History of Medicine, &c., vol. i. p. 277.
† Wood's Zoology, vol. iii.

Xenop. unicolor. None are classified under genera found in Australia or in Africa. Whether any varieties exist in South America is not known.

FOURTH GENUS—UROPELTIDÆ. *I. Muller.*

"Rough tails." Tail extremely short, truncated, scarcely tapering; generally terminating in a rough, naked disk, or covered with carinated (keeled) scales.

East India species are:

No. 1. Uropeltis Rhinophis. *Hemprich.*
2. " R. oxyrhynchus.
3. " R. punctatus.
4. " R. Philippinus.
5. " R. Trevelyanus.
6. " R. sanguineus.
7. " R. Blythii.
8. " R. Pulneyensis.
9. " Grandis. *Kelaart.* A rare variety.
10. " Plecturus. *Dum. and Bibr.*
11. " Pl. Perotettii.
12. " Pl. Güntherii.
13. " Mellanophidium. *Dum. and Bibr.*
14. " Mell. Wynandense. *Beddome.*
15. " Silybura. *Peters.*
16. " Syl. macrolepis.
17. " " Beddomii.
18. " " ocellata.
19. " " Elliottii.
20. " " bicatenata.
21. " " Shortii.
22. " " brevis.

Africa and Australia furnish no varieties which have as yet been classified.

In South America there are several varieties.

Fifth Genus—CALAMARIDÆ. (*Dwarf Snakes.*)

Ventral shields distinct; two occipitals, tail tapering.

Six species, comprising twelve varieties.

Several varieties are found in India, and many more in South America, but the following are such species as have been already classified, viz.:

No. 1.	Calamaridæ	Calamarea.	5 varieties.
2.	"	macrocalamus.	1 variety.
3.	"	oxycalamus.	1 "
4.	"	geophis.	1 "
5.	"	aspidura.	3 varieties.
6.	"	hoplocercus.	1 variety.

Dr. Russell describes a variety found in East India, which is also mentioned by Seba, viz.:

Abdominal scuta 151, subcaudal squamæ, 93 = 244. Length 13″, tail alone 4″. Called "Dooblee." Head a little broader than the neck, oblong-ovate, compressed, 10 laminæ; mouth large, no fangs, marginal row of teeth in upper jaw; eyes large; nostrils small, close; trunk round; tail very tapering, slender, and terminates in a long sharp point. Scales on the back ovate, carinate, on the other part smooth; color of head light brown, trunk freckled with dusky yellow spots; scuta yellowish-white; most of them with black margins.

Another variety described by Russell is called, in Tranquebar, Neer Pamboo; in Calcutta, Dooblee.

Abdominal scuta 146, subcaudal squamæ, 82 = 228. Length 1′ 2¼″, tail 3⅝″. Color dusky lead, with numerous black dots on back and sides; abdomen, and under part of tail white.

A third variety, called in Tranquebar, Neer Pamboo.

Abdominal scuta 143, subcaudal squamæ 83 = 226. Length 14'', tail 3¼''. Color less dark than preceding, and has spots of dusky yellow interspersed with black dots.

This and the preceding varieties have two spots on the last pair of laminæ on the occiput, which are not seen in the larger individuals. The details of the *Ourdia* answer for both these varieties.

OBSERVATIONS.

Two other individuals were brought to Dr. Russell with the name *Neer Pamboo*, but the one was *Bokadam* and the other *Chittee*. The tail in these varieties is exceedingly thin and fragile, and difficult to preserve entire.

SIXTH GENUS—OLIGODORITIDÆ. *Boie*.

Body rather rigid, covered with rounded, smooth scales; abdominal scuta developed; *head* short, not distinct from neck; maxillary teeth few in number, the last one enlarged, not grooved; no fangs.

These are small snakes, said by some naturalists to be peculiar to the East Indies, but many varieties are known to exist in South America.

The genus is composed of two species and twenty-eight known varieties, viz.:

    No. 1. O. Oligodon,   12 varieties.
       2. O. Simotes,   16   "
                                  28   "

Russell describes two individuals as follows, viz.: One called Shilay Pamboo.

Abdominal squamæ 207, subcaudal squamæ 6 = 213. Length 9'' (inches).

The other is Anguis maculata, of Linnæus. This, as well as the preceding, has no fangs.

Abdominal squamæ 194, subcaudal squamæ 6 = 200. Length 10''.

SEVENTH GENUS—COLUBRIDÆ.

Is composed of four species and thirty-four varieties, viz.: the varieties being divided into seventy-seven sub-varieties. The species are:

1. Coronellina (ground Colubers).
2. Colubrina (true Colubers).
3. Dryadina (bush Colubers).
4. Natricina (fresh-water Colubers).

The first species is composed of the following varieties:

No. 1. Col. Cor. *Ablabes.*   8 sub-varieties.
    2.   "   *Cyclophes.*   5   "
    3.   "   *Odontomus.*   3   "
    4.   "   *Nymphophidium.*   1 sub-variety.
    5.   "   *Caronella.*   1   "
    6.   "   *Coluber.*   3 sub-varieties.
    7.   "   *getulus:* the Thunder Snake, King Snake, or Chain Snake, is an aggressive variety, although it has no fangs. It has been known to kill and swallow a Rattle-snake.
    8.   "   *quadrivittatus:* 1 sub-variety; a North American variety; indulges a propensity in stealing chickens from their roost, and swallowing them.
    9.   "   *eximus:* 1 sub-variety; North American variety is called the Milk or House Snake.
   10.   "   *Esculapii:* 1 sub-variety; of India, Australia, and Japan; was represented by the ancients as twined around the staff of Æsculapius, and also the Caduceus of Mercury.

No. 11. *Col. Cor. canus:* 3 sub-varieties; is found in South Africa.

12. " *coronella:* 1 sub-variety; multimaculata; also found in South Africa.

The Australian species are:

No. 13. *Col. cor. cornella Australis:* 1 sub-variety.

14. " *mycterizans* (Linn. 1 sub-variety; called Passeriki Pam, Pastiletti, by Russell.

Abdominal scuta 178, subcaudal squamæ 166 = 344. Length 4′ 6″ of a large specimen, diameter ½″. *Head* much broader than the neck, oblong, depressed above, rounded on the sides, compressed and contracted at the eyes. Rostrum long, straight, sharp-pointed, resembling the beak of a bird, with a small, soft, obtuse, reflex process at the extremity. *Occiput* covered with eleven laminæ, and suborbicular, imbricate scales; *teeth* slender, reflex, three rows in upper jaw. *Eyes* lateral, large, prominent; *nostrils* small; trunk slightly triangular; *abdomen*, flattish; *spine* slightly carinated; *scales* linear, lanceolate, loosely set on the neck and front part of the trunk, but the rest closely imbricate; *tail* long, slender, and covered with exceedingly minute scales at the point. *Colors:* the head appears to be covered with green velvet with a streak of yellow on each cheek; the rest of the body is of a yellowish grass green when quiet, but when provoked the body swells, and the interstitial white between the scales comes out, while the yellow assumes a changeable hue. From the throat to the anus there runs on each side of the body a yellowish-white fillet, which assumes a darker shade near the anus, and terminates near the tip of the tail. The scuta and squamæ are of a light yellowish-green.

OBSERVATIONS.

This variety is very common about Vizagapatam and in the Circars, and is said to abound in the Carnatic. Found often in the trees, and the natives say it drops down upon passers-by. Its poison produced no noxious effects in chickens. The *C. mycterizans* of Linnæus is marked venomous; this is an error pointed out by Dr. Gray.

No. 15. *Col. cor. stolatus* (Linn.). 1 sub-variety. Called Wanna Pam, by Russell.

Abdominal scuta 143, subcaudal squamæ 70 = 213. Length 1′8″, diameter ¾″; seldom found to exceed this size. Head somewhat broader than the neck, rather short, obtuse ovate, depressed, covered with nine laminæ, the large posterior pair oblique, oblong, semi-cordate, with two or three small laminæ on each side. Mouth large, three rows of teeth in the upper jaw. Eyes large. Trunk round, invested with oval, thick-set carinated scales. Color of head and neck a very dark green; cheeks and throat yellow; on the neck are two blackish bands, from which a yellowish-brown fillet runs on each side along the back and part of the tail. This fillet is marked, between the fillets on the body, with broader, transverse, whitish bands, and between the fillets and the abdomen, with waving, interrupted, longitudinal white lines. The ground color between these approaches to black. From just above the anus to the tip of the tail the ground color is black, marked by fillets of a uniform color. Scuta and subcaudal squamæ are of a dull pearl color, the former having a black dot on each side.

OBSERVATIONS.

No poison-organs could be found in this variety.

The *C. stolatus* is marked poisonous by Linnæus, but, as Dr. Gray observes, this is an error. Two specimens from Ganjam were called—

Neerogady; scuta 146, subcaudal squamæ 77 = 223; and Neerogady; scuta 147, subcaudal squamæ 71 = 218.

No. 16. *Col. cor. lineatus* (Linn.). 1 sub-variety. Called Jeri Potoo, by Russell.

Abdominal scuta 176, subcaudal squamæ 88 = 264. Called Condenarouse in Ganjam, and much feared. Length 2′ 4½″, diameter 1″. Head broader than neck, oblong, ovate, depressed above, compressed toward the rostrum, covered with nine laminæ. Mouth medium size; lower jaw projects beyond the upper; three rows of teeth in the upper jaw, two front teeth unusually long. Eyes large, lateral, oval. Nostrils small. Trunk round, diminishing regularly from centre towards the neck and to the tail, covered with oblong, oval, smooth scales. Color of head light brown, a yellow streak behind each eye. Trunk and tail are striped with seven longitudinal fillets, of which the centre and the two outside ones are of a darker brown, of a greenish cast, and broader than the others; one on each side of the centre brown fillet is almost black, and the adjacent one on each side is of a greenish yellow. The scuta and one-half the squamæ nearest the anus are of a straw color; a small darkish-green thread runs along the sides of the scuta and squamæ to near the end of the tail.

No. 17. *Col. cor. mucosus* (Linn.). 1 sub-variety, called (*Russell*) Jeri Potoo in Vizagapatam.

Abdominal scuta 199, subcaudal squamæ 121 = 320. Length 5′ 4″, diameter 1¼″. *Head* proportionately small and narrow, acute ovate, depressed, but compressed toward the rostrum; nine principal laminæ, but the sides of the occiput are covered with several smaller ones. Upper jaw slightly divided, has three rows of teeth. *Eyes* lateral, large, prominent. *Nostrils* small, gaping. *Trunk:* the neck is covered with small oval, smooth, imbricate scales; back carinated; sides compressed; scales subrhomboidal, but four rows on the upper part of the back are either carinated or striated; *tail* small, taper, sharp-pointed, nineteen inches long. *Color* of cheeks and sides of throat a pale flesh or whitish, streaked transversely with black lines. The jugular scuta are yellowish white, each having a blackish spot on each side. Head, neck, and part of the trunk are a dull yellowish olive color, while the remainder of the trunk and tail has the color somewhat lighter, and is variegated by transverse black lines and spots joined together, and which become blacker as they approach the tail. Half the abdominal scuta are of a dull white, strewed with dusky spots, but the lower edge of each scale is purplish black, and all the scales on the tail have black edges. Subcaudal squamæ are greenish yellow with black edges.

OBSERVATIONS.

Bites of this variety in chickens showed no bad results.

No. 18. *Col. cor. nasutus.* 1 sub-variety (Shaw). Aspides, (Seba).

Length 4′ 11″; length of *tail* 1′ 8″

No. 19. *Col. cor. Austriaca;* has been found in England.

20. *Col. cor. cana.* Schwartze Schlange, or Black Snake,

occurs in Africa, and also in the United States of North America.

The second species, *Colubrina*, is composed of 6 varieties, made up of 16 sub-varieties, as follows, viz.:

No. 1. Col. C. Elaphis.   3 sub-varieties.
     2.   "   Compsosoma. 4   "
     3.   "   Cynophis.   2   "
     4.   "   Ptyas.   2   "
     5.   "   Henelaphis. 1 sub-variety.
     6.   "   Zamenis.   4 sub-varieties.

The greater part of these are found in India; many others occur in South America; but, as yet, none of them are classified.

The third species, *Dryadina*, is composed of 4 varieties, made up of 7 sub-varieties, as follows, viz.:

No. 1. Col. Dryadina zaocys.   4 sub-varieties.
     2.   "   herpatoreas. 1 sub-variety.
     3.   "   guttatus. 1 sub-variety, or Corn Snake; abounds in North America, and is from 5 to 6 feet long.
     4.   "   arboreum virides. 1 sub-variety; is the North American Green Snake.

The fourth species, *Natricina*, is composed of 4 varieties, made up of 17 sub-varieties, as follows, viz.:

No. 1. Col. natricinus tropidonotus. 13 sub-varieties.
     2.   "   atretium.   2   "   Chittee R.
     3.   "   xenochrophis. 1 sub-variety.
     4.   "   prymnomiodon. 1   "
     5. ***Col. natr.*** *atretium*, has a sub-variety, called Chittee in India, described as follows, viz.:

Abdominal scuta 154, subcaudal squamæ 67 = 221. Length 1'7''; tail

3¾″; *head* oblong-oval, compressed on the sides, a little broader than the neck, obtuse, flat on the crown; ten laminæ, with a singular triangular-shaped lamina between the nostrils; *mouth* of middle size, three rows of teeth in upper jaw; no fangs; *eyes* large; nostrils small, close; *trunk* round; tail tapers rapidly, and terminates in a sharp point; scales smooth, ovate, small; *color*, crown and upper part of body a bluish clay; body variegated with dusky spots; belly a tawny buff.

No. 6. The Col. natr. trop. picturatus is a sub-variety, found in Australia.

No. 7. *Col. natr. trop. natrix*, is the Grass or Ring Snake, common in England.

Eighth Genus—HOMALOPSIDÆ.

Fresh-water snakes; composed of 13 species, made up of 19 varieties.

The distinguishing mark for this genus is that the openings of the nostrils are on the top of the snout. Günther gives 8 species, as follows, viz.:

No. 1. Homalopsis (Tatta Pam).   1 sub-variety.
    2. H. fordonia.   1   "
    3. H. cantoria.   1   "
    4. H. cerberus (Karoo Bokadam). 1   "
    5. H. hypsirhina (H. hyps. aer.). 5 sub-varieties.
    6. H. ferania.   1 sub-variety.
    7. H. hipistes.   1   "
    8. H. herpeton (H. tentaculus). 1   "

Russell describes an East India variety, called Tatta Pam in Vizagapatam, and Ang. Scytalæ (Linn.), as follows, viz.:

Length, 19½″, tail 2″ long; is flat and eel-shaped, like that of the *Nalla Wahlogillee Pam*. *Head*, hardly broader than the neck, small, roundish, obtuse; 12 laminæ; *mouth* narrow; three rows of teeth in upper jaw; *eyes* lateral; nostrils vertical, gaping; *trunk*, neck round and smooth; *scales* small, ovate, imbricate; *back* carinated; *sides* sloping, belly roundish, scales orbicular, close, not imbricated on trunk, tail, and belly; thickest part of the body at and near the anus, and tail double the thickness of the neck; *color*, head black, trunk and tail black, but on the sides are 68 yellowish-white conical spots, with points toward the ridge of the back, some of which, on the neck and near the tail, join their points together; tip of tail not spotted.

OBSERVATIONS.

This specimen was found on the sea-beach in Vizagapatam; but, being put into salt water, with an idea to preserve it, at the expiration of a few minutes it died.

Species No. 4 has a variety found in the East Indies, called Karoo Bakadam in Ganjam, and described by Russell, as follows, viz.:

Length, 3½ to 4 ft.; diam. of neck 3½″, and body 4½″, in its thickest part; tail 8″ long, has an obtuse point. *Head* slightly broader than neck, but small proportionately, slightly convex above, compressed at the sides, projecting into a short, subtruncate snout; *eyes* placed higher up than the other varieties, protuberant, and each one forms the centre of a remarkable ring, formed by triangular laminæ. The snout is covered with several small laminæ, but the remaining part of the head is covered with suborbicular, carinated scales. *Three rows* of teeth in the upper jaw; no fangs; *trunk* thick, round, covered with large, carinated, broad, oval, imbricated, scales. *Color* of the head, back of the lamina, almost black, *trunk, tail.* and *snout,* a dark gray; throat, belly, and under part of the tail, are a dusky yellow.

Another East India variety is Velar Sawa (*H. hyps. aer.*); it is said to attain a length of 30 feet.

Herp. tentaculatus, another East India variety, has two tentacula-like projections on the snout, which are two or three inches in length.

Herp. nasutus (nobis), *Langaha nasuta* (Wood), is a Madagascar variety, said to have a singular long snout, like a proboscis.

No. 9. *H. chersydrus granulatus* (Wood), is found in Africa.

No. 10. *H. herp. crista galli* (the Cockscomb Langaha), a Madagascar variety, bears an appendage on its head like a cock's comb.

No. 11. The *H. eunectes marina* (Wood), is a Brazilian variety, called Traga-venado, or deer-swallower, or anaconda. This should be called *H. eunectes python*. It is found in great abundance in the pampas of the great valleys of South America. The Boa Constrictor of Brazil, called Boiguaeu, differs little, if at all, from the preceding.

Krefft gives, No. 12, *H. cerberus australis*, and No. 13, *H. myron Richardsonii*, as the Australian varieties. South American varieties are exceedingly numerous, but not classified.

Ninth Genus—PSAMMOPHIDÆ. *Boie.*

Desert Snakes, peculiar to the great deserts, and dry arid places.

Body elongate or stout; pupil round or vertical; *loreal* region concave; one of the four anterior maxillary teeth is enlarged, the last tooth grooved.

First Species—*Ps. condonarus*, Russell; for description see *Col. cor. lineatus*, page 31.

Second Species—*Ps. pulverulentus*, Günther.

Third Species—*Ps. psammophylax rhombeatus* (Schaapsticker) is a New Holland variety. No varieties have been classified as yet under Australian species.

Tenth Genus—DENDROPHIDÆ.

Tree Snakes.

Body and tail much compressed, or very slender; *head* elongate, snout obtuse or rounded, pupil round; no fanglike tooth in front.

The following species occur in India, viz.:
- No. 1. D. gonyosoma, *Wagl.*
- 2. D. phyllophis, *Günther.*
- 3. D. dendrophis.
- 4. D. crysopelea.

South Africa furnishes three species, viz.:
- No. 5. D. semivariegata.   2 varieties.
- 6. D. natalensis.   1 variety.
- 7. D. albovariegata.   1 "

                                      4 varieties.

Australia furnishes two species, viz.:
- No. 8. D. punctulata, 1 variety (Green Tree Snake).
- 9. D. calligastra, 1 variety (Northern Tree Snake).
- 10. D. multimaculatus (nobis), is an East India species, known as Goobra in Hyderabad.

Length 3' 2", diameter ¼", length of tail one foot, very slender, and terminates in a sharp point. *Head* broader than the neck, long, ovate, depressed, and covered with nine laminæ; occiput is covered with small orbicular scales; *mouth* large; *three rows* of teeth in the upper jaw; no *fangs; eyes* large, oval; *nostrils* large, open; *trunk* covered with small, linear, oval scales; middle of back somewhat depressed, its scales are obovate; two rows of oval scales on the sides of the scuta; *color* of head and trunk a dark sepia brown, but a shade of lighter color runs along the ridge of the spine; towards the scuta the abdominal oval scales are whitish, spotted with black, up to the middle of trunk, thence to the end of the tail they are of the same cineritious color as the scuta; some of the dark-brown scales on the sides are tipped with black above, and with a beautiful azure blue below.

OBSERVATIONS.

This variety is said to grow much larger, and to be found mostly on trees; not known to be poisonous.

Eleventh Genus—DRYIOPHIDÆ.

Whip Snakes.

Body generally exceedingly slender; head very long, with tapering snout; pupil linear, horizontal; last maxillary tooth grooved.

East India varieties are as follows, viz.:

No. 1. Dry. tropidococcyx. 1 variety.
    2. Dry. tragops. 3 varieties.
    3. Dry. passerita. 2 "
    4. Is a South American variety, called Culebra Bejuco; a very great number of these snakes are found on the plains of Casanáre and San Martin.
    5. Dry. acuminata, Golden Tree Snake, is a Mexican variety.
    6. Dry. herpetodryas flagelliformis (nobis), Coachwhip Snake, is a harmless North American species.
    7. Dry. philodryas viridissimus, Emerald Whip Snake, is from Brazil, a harmless variety, and of a beautiful color.

Making in all 7 species, composed of 10 varieties.

Twelfth Genus—DIPSADIDÆ.

Nocturnal Tree Snakes; with a vertical pupil; body and base of tail much compressed, head subtriangular, broad behind, very distinct from neck, with short snout; loreal region flat.

East India species are:

No. 1. Dips. cynodon. 1 variety.
    2. " forsteni. 1 "
    3. " boops. 1 "
    4. " dendrophila. 1 "
    5. " bubalina. 1 "

No. 6. Dips. multimaculata.   1 variety.
    7.   "   trigonata.   1   "
    8.   "   multifasciata.   1   "
    9.   "   gookool.   1   "
   10.   "   ceylonensis.   1   "
   11.   "   eudipsas cynodon, is an Asiatic species.
   12.   "   inornatus, is an African species.

Australia furnishes one species, viz.:
   13. Dips. fusca, Brown Tree Snake.

Making in all 13 species, composed of 13 varieties.
Many species are known to exist in South America.

Thirteenth Genus—LYCODONTIDÆ.

Ground Snakes, of which few species are as yet classified.

Maxillary teeth, with a fanglike tooth in front, but without elongate posterior teeth.

East India species are, viz.:
No. 1. L. lycodon.   5 varieties.
   2. L. tetragonosma.   2   "
   3. L. leptorhytaon.   1 variety.
   4. L. ophites.   2 varieties.
   5. L. cercaspis.   11   "

South Africa furnishes three varieties:
No. 6. L. capensis.   1 variety.
   7. L. geometricus.   1   "
   8. L. guttatus.   1   "

Making in all 8 species, composed of 24 varieties. Varieties existing in South America are almost innumerable.

Fourteenth Genus—AMBLYCEPHALIDÆ.

Blunt Heads.

Body flexible; *ventral shields* developed; *head* thick, very distinct from neck; no mental groove; *maxillary teeth* very small, few in number.

East India species are as follows, viz.:

No. 1. A. amblycephalus, 1 variety.
    2. A. pareas, 3 varieties.
                      ―
                      4

Making 2 species, composed of 4 varieties; none other than these have as yet been classified.

Fifteenth Genus—PYTHONIDÆ.

Rock Snakes.

Body cylindrical, flexible; interior maxillary teeth unequal in length; none of the hinder teeth enlarged; rudiments of hind limbs present; tail prehensile.

Several species occur in East India, viz.:

No. 1. *P. Python* (Daud). 2 varieties.
    2. *P. Natalensis* (Port Natal Rock Snake) is said to attain a length of twenty-five feet. This is an African species.
    3. *P. aboma* (nobis). Aboma, or Ringed Boa of Mexico, was formerly worshipped by the natives on account of its destructive powers.
    4. *P. morelia spilotes* (Diamond Snake) is a handsome variety.
       P. morelia variegata (Carpet Snake) is closely allied to the preceding. These are varieties found in Australia.
    5. *P. molurus* (Common Rock Snake) has a spur on either side of the anus. This is an East Indian variety.
    6. *P. aspidiotes melanocephalus* (Black-headed Snake).
    7. *P. liasis childrenii* (Children's Rock Snake).
       P. liasis olivacea (Olive-green Rock Snake).

No. 8. *P. nardoa Gilbertii* (Gilbert's Rock Snake). These are varieties found in Australia.

9. *P. lazuli* (nobis), called Pedda Poda.

Abdominal scuta 252, subcaudal squamæ 62 = 314. Length 2' 9'', diameter 1¼''. *Tail* round, tapers regularly, ending in a sharp point, length a little more than four inches. *Head* long, broader than the neck, hastate, obtuse, depressed, covered with twelve principal laminæ, besides several smaller, disposed in a star shape around the posterior ones. The ventral lamina between the eyes is divided throughout the middle,—a very unusual occurrence. Occiput is covered with very small smooth orbicular scales. *Mouth* wide, periphery rather thick, and covered with oblong transverse scales; three rows of teeth in upper jaw. *Eyes* lateral, small. *Trunk* round, closely set with minute, smooth, round imbricate scales. Three rows of scales next to the scuta larger than the others, oval, acuminate. *Color:* upper part of the head flesh color; rostrum ashes color; a little streak of flesh color runs obliquely along on each side of the neck, and a short, narrow, wedge-shaped stripe, of the same color, divides a large brown mark on the occiput. Trunk and tail ashes color, variegated with about thirty large, broad brown maculæ, edged with black, of different forms and size. Sides spotted with similar smaller maculæ, most of which are whitish in the middle. Scuta remarkably small and narrow, whitish, with reddish margin around posterior edge of anus. Squamæ small. Upper part of tail singularly variegated in white and black, the latter color in long broadish streaks.

OBSERVATIONS.

Experiments made on fowls did not result fatally in any case, although this variety has a very formidable look.

No. 10. *P. bimaculata* (nobis), called likewise Pedda Poda.

Abdominal scuta 252, subcaudal squamæ 64 = 316. Length 7' 2½'', diameter 2¾''. *Tail* short, round, tapers to a sharp point. *Head* slightly broader than neck; oblong rostrum; rounded, depressed above, sub-compressed toward the rostrum; eleven principal laminæ are contiguous, but these are surrounded with numerous small ones; occiput covered with smooth ovate scales. *Mouth* large; three rows of teeth in upper jaw. *Eyes* not lateral, prominent. *Nostrils* small. *Trunk* round; scales small, smooth, ovate imbricate; two or three rows next to the scuta are larger; scuta and squamæ are acuminated and exceedingly small. *Color* of head dark ashes, with a dark-brown oblique macula behind each ear; neck, trunk, and tail variegated with large dark spots with irregular forms, edged with black on a light-brown ground; spots on tail are somewhat lighter in color than those on the trunk; scuta and squamæ of a dusky yellow.

OBSERVATIONS.

This specimen came from Ganjam.

No. 11. *P. albo-maculata* (nobis), called Pedda Poda.

Abdominal scuta 256, subcaudal squamæ 69 = 325. Length 6′, diameter 2¼″. *Head* a little broader than the neck, depressed above, oblong; forward of the eyes contracted, compressed, subtruncate; crown and front part covered with many laminæ, but occiput is covered with smooth ovate scales. *Mouth* very wide; three rows of teeth in upper jaw. *Eyes* medium size, lateral, protuberant. *Nostrils* small, gaping. *Trunk* covered with smooth ovate imbricate scales, except two rows on the belly, which are orbicular; scuta are oblong, narrow, acuminate at each end; subcaudal squamæ are oval, but pointed. Above the anus and on a line with the penultimate scuta, there is on each side a small curved horn or spur, turned outwards *Color* universally whitish, variegated with long, brown, irregular-shaped spots, edged with black. There is a long diagonal dark streak behind each eye, and the first dark spot on the occiput is divided, half way down its middle, with a whitish streak. The squamæ and some spots on the under side of the tail are of a dusky slate color.

OBSERVATIONS.

The last three varieties (Nos. 9, 10, and 11), belong to what the natives in Ganjam call Rock snakes, and are not considered venomous. Specimens of No. 11 have been seen 10′ long; this variety is (identical with the Bora of Bengal), called Dussery Pamboo, which has claws identical with this one, and the snake men affirm it is necessary at times to clip them.

No. 12. *P. bora* (nobis) (Bora) is described by Dr. Russell, as follows, viz. :

ERYCIDÆ. 43

Abdominal scuta 265, subcaudal squamæ 36, subcaudal scuta 28, subcaudal squamæ 3 = 67 = 332. *Length* 4' 10''; *diameter* nearly 2''; *tail* round, short, 7½'' long, pointed. *Head* a little broader than the neck, compressed, very obtuse; *occiput* covered with small ovate scales, ten laminæ; *mouth* wide, jaws of equal length; *three* rows of teeth in the upper jaw, those in lower jaw very long and large; *eyes* large, lateral; nostrils close together; *trunk* round, covered with small, smooth, ovate scales, closely imbricate; on each side of the scuta are two rows of larger ones; scuta remarkably short; on each side of the anus is a horny spur, about ¼ of an inch long (like the Pedda Poda, No. 11). The scuta at the end of the tail are singular. *Color:* brown predominates; all along the back are large, roundish spots, light brown in the centre, edges yellowish brown; sides are variegated by brown spots on a whitish ground, which brightens near the belly; scuta pearl white.

OBSERVATIONS.

It is pretended that the bite of the Bora does not prove mortal till after 10 or 12 days, but that eruptions in different parts of the body soon follow the introduction of the poison into the blood. Nothing in the snake seems to warrant a belief that it is poisonous.

Innumerable varieties and species of this genus abound in South America, some individuals attaining to an enormous size. One was killed on the head waters of the River Amazon 30 feet in length, and more than 3 feet in diameter; and an English nobleman saw one, just after it was killed, on the island of Manilla, which measured over 31 feet long, and more than 39 inches in diameter.

The number of species, at present classified, is 13, composed of 16 varieties.

SIXTEENTH GENUS—ERYCIDÆ.

Sand Snakes. Tail very short, not prehensile. East Indian species are:

   No. 1. E. eryx.   1 variety.
      2. E. gongylophis.   1   "
      3. E. cursoria.   1   "

Several species, not yet classified, occur in South America, and also in Mexico.

Seventeenth Genus—ACROCHORDIDÆ.

Wart Snakes. Some species are not venomous, and others are known to be exceedingly so. East India species are, viz.:

No. 1. Acrochordon xenodermis.   1 variety.
    2. Acro. javanicus, 8 ft. long.   1   " (Java).
    3. Acro. chersydrus.   1   "

One of the preceding varieties, called *Oular carron*, is often eaten by the natives in India, and its flesh esteemed very delicious.

No. 4. *Acro chocoe* (nobis), Verrugosa, is a species in great abundance in the forests of the River Atrato, in the United States of Colombia. Its poison is known to be very deadly.

According to Du Chaillu, it is also found in Africa. Its generic description is: *body* covered with small, wart-like, non-imbricate, scales; tail prehensile; *length* from 3' to 7'; *diam.* $\frac{3}{4}''$ to 3''; dorsum covered with irregular, wart-like projections from head to tail, without scales; *color* variegated, brown, dark-colored, and cream-colored spots, some of the dark ones approaching to a greenish hue. Immediately after the death of the snake, a thick, milk-like liquid exudes from these warts, which, if applied to the skin of man or beast, produces a wellnigh incurable ulcer. It has large fangs, and its buccal parts are precisely similar to those of the Crotalus horridus. This serpent is the most feared by the natives of all those found in the Chocó region; the "Curers" say that its bite produces death, frequently in from 2 to 3 hours; and that the first symptoms which the poison develops are, lethargy, trembling of the muscles in the whole body, a flow of blood from the pores of the skin, eyes bloodshot, and loosening of the hair, followed by a distortion of the features. What effects the poison does actually produce can only be determined by experiment. Some of its effects, in cases of bites, are the following: Soon after the introduction of the poison into the blood, the skin of the surface of the bitten limb is thickly covered with small vesicles, filled with an ichorous liquid; when these have attained the size of a grain of barley they burst, leaving a small sore, which soon increases in size, preserving its funnel shape; these continue sloughing away, until they unite one with another, and thus destroy all the fleshy substance down to the bone. This is accompanied with intense throbbing pains up the limb. In cases where the poison does not cause death, on account of its deadly principle not being fully developed, it almost invariably produces these funnel-shaped ulcers. The Indians cure the bite of this serpent with a tincture of a plant called cock-flower (flor de gallo), of the nature and virtues of which particular mention is made in another place. How specific this is, however, is not positively known.

OPHIDIA VENENOSI.

The Ophidia Venenosi comprise the following genera:

    No. 18. Elapidæ.
         19. Hydrophidæ.
         20. Crotalidæ.
         21. Viperidæ.

Generic distinctions are, viz.:
The existence of fangs or teeth, with their accompanying secretory apparatus and receptacle for the poison.

The number and arrangement of the non-imbricated shields of the head are as follows:

Rostral.	Ocular anterior (or preorbital).
Anterior frontal.	Ocular posterior (or postorbital).
Posterior.	Upper labials.
Vertical.	Temporals.
Supraciliary.	Mental.
Occipital.	Lower labials.
Nasals.	Chin shields.
Loreal.	

The heart has three cavities, one ventricle, and two auricles; the urinary bladder is wanting. They are all carnivorous and swallow their prey entire; are *oviparous, ovo-viviparous,* or *viviparous,* and hæmatocrya or cold-blooded. Their blood is rich in solid constituents, and has red corpuscles of an elliptical shape, flattened, biconvex, and smaller than in other reptilia. Of the 21 genera already classified, 4 only have been considered poisonous; but in South America varieties occur under the genera Calamiris, Homalopsis, and Acrochordus, which are known beyond a doubt to be poisonous; and some species, which must be classified under the genera Python and Typhlope, found in India, are also venomous.

This divides the 21 genera into

Non-venomous,	12 genera.
Venomous,	9 genera.

The most deadly of these, viz., Ophiophagus elaps and Naja tripudians belong exclusively to India, whilst the Viperidæ are represented by the *Daboia Russellii,* and the *Crotalidæ* by the *Trimeresuri.* The latter genus is less venomous than the *Crot. horridus* of North America; the *Crot. cascabellus;* the

Crot. fasciatus, or the *Craspedocephalus Brasiliensis*, called in Brazil, Jararacea; but there are Viperidæ in South America, e. g., *Vip. pseudechis carinata*, *Vip. calamaris venenosi*, as well as the *Trigonocephalo lachesis*, the *Echis variegata*, and *Acrochordon chocoe*, whose poisons are little less venomous than that of the *Naja tripudians*. Some varieties of Ophidians have fangs, though not venomous; and others, although venomous, possess none.

In cases of bites, the pterygoid, maxillary or palatine teeth seldom or never leave their marks upon the skin, and sometimes it is very difficult to distinguish even the small incisions made by the fangs. When these wounds are distinct they show oftentimes indications of three fang-wounds on one side, and three or two on the other, all close together; but in these casual cases *the forward fang* alone is in communication with the poison-bladder.

In all cases where *more than one* fang occurs on either side this is *invariably* the result of casualty. Snakes often have their fangs loosened or broken off. In the former case a spare fang comes forward and attaches itself to the base of the bone to which the loosened fang was fixed. Twenty-four hours suffices for it to solder itself firmly to its base; and the loosened fang becoming resoldered at the broken part leaves thus two fangs on the one side, while on the other side only one fang is found.

Where the two fangs are, *only the forward one* is in communication with the poison-bladder; and thus it happens that individuals are found which have as many as four fangs on one side, and only two or three on the other, but this unnatural condition is the result of casualty, as explained above. The usual number of spare fangs found in the fang-sac is *seven* on each side, when the snake is fully grown, and has as yet lost no fang by accident. All snakes with fangs have the arrangement of their buccal parts identical with those of the *Crotalus horridus.*

Those not provided with fangs have the same parts identical with those of the *Elaps corallinus.*

Large snakes, the Pythonidæ for example, have very large and long fanglike teeth, in proportion to their size, but only the venomous varieties have the poison-bladder, or the apparatus for injecting it. Some naturalists have given drawings showing the shapes of the fangs and their attachments, in which the base of the fang is *turned backwards.* Of more than one thousand specimens examined carefully, and with scrutiny, in South America, I never have found a venomous snake whose fangs had the above shape; they are invariably attached on the line of curvature of the fang, and their base *never curves backwards.*

Eighteenth Genus—ELAPIDÆ.

Divided into two species, viz.:*

  No. 1. *Hajidæ,* snakes with hoods; and
    2. *H. elaps,* snakes without hoods.

Hajidæ. { 1. Haja, the smaller variety.
    2. Ophiophagus, is the largest poisonous snake known.

*First Species—*HAJA. (*Cobra di Capello.*)

Coluber Haja, *Linn.*   Haja larvata, *Cantor.*
Haja lutescens, *Cantor.*   Haja atra,   "
Haja tripudians, *Günther; Gray.* Haja Kaouthiah, "

In Bengal, Cobras with the spectacle-mark are called Go-

---

* Thanatophidia.

kurrahs; those without it, but with another mark, or ocellus, on the hood, are called Keautiahs.

No. 1. Kala, or black.
 2. Koyah, or black and white.
 3. Gomunah, or wheat-colored.
 4. Puddah, yellow-colored.
 5. Dudiah, whitish-colored.

No. 6. Tentuliah, tamarind-seed-colored.
 7. Kurrees, earth-colored.
 8. Tameshur, copper-colored.
 9. Puddun-nag, golden-colored.

The most common varieties about Calcutta are the 2d, 3d, and 7th. Of the Keautiahs there are also 9 varieties, viz.:

No. 10. Kala.
 11. Tentuliah.
 12. Kurrees.
 13. Sonera, golden-colored.
 14. Dudiah.
 15. Bans Buniah, white or black mottled.

No. 16. Giribungha, brownish-colored.
 17. Koyal.
 18. Sankha Mookhi, black and yellow.

Nos. 10, 11, and 15 are most common in the vicinity of Calcutta. In Hindostan the Cobra is called Kála Sámp, Nág Sámp in Vizagapatam, and Bengal, Nagoo.

Russell describes the following varieties, viz.:

No. 19, called *Chinta Nagoo*, Coluber Naja, *Linn.*

Length 4 ft.; diameter more than 1″. *Head* hardly broader than the neck, short, broad ovate, obtuse; crown depressed from the eyes, contracted, compressed and declining toward the rostrum; laminæ ten; scales on the occiput small, orbicular, and oval; *mouth* large; lower jaw has few teeth, upper jaw two rows; two fangs; eyes small, prominent; nostrils close together, gaping.

When the reptile is at rest the neck is little wider or thicker than the head, but when provoked the loose skin is extended in a peculiar manner so as to form what is called a hood; this fact gives the name to the species, *Cobra di Capello*, meaning "a monk's cowl or hood." The peculiar spectacle-shaped mark on the back of the hood is partly formed by the color of the interstitial skin, but the color of the scales also contributes to give it this shape. *Trunk* round, scales rather small, oval, polished, contiguous, and hardly imbricate except on the hinder part of the tail, but a row on each side of the belly are larger, ovate, and imbricate; tail tapers gradually to a sharp, horny point. *Color* yellowish light brown; interstitial skin white, and the edges of some of the scales white. In certain positions the scales reflect a faint bluish-ash color; abdominal scuta are very long, but the subcaudal squamæ are hexagonal; both are of dull white, and freckled with dusky spots. The colors of the spectacle-mark are bright, but the orange tint of the interstitial skin is not so distinct as in some other varieties.

No. 20. *Arege Nagoo.* (*Arege* means small grains that the horses eat.)

Abdominal scuta 189, subcaudal squamæ 60 = 249.

The spectacle-mark has a brown spot in its centres, and five of the cervical squamæ are notably darker than the rest. There are also two dark-brown spots, one on either side of the cervix, which forms when extended the inside of the hood.

No. 21. *Coodum Nagoo.* (*Coodum* signifies wheat.)

Abdominal scuta 187, subcaudal squamæ 57 = 244.

This variety is rather darker than the others, and the orange color more inclined to yellow. The principal and distinguishing mark being an oblong curved loop instead of the spectacle-mark, the skin in the centre of the loop being white.

No. 22. *Sankoo Nagoo.* (*Sankoo* is the name of a shell used for glazing paper.)

Abdominal scuta 183, subcaudal squamæ 56 = 239.

This variety is rare, and is distinguished by a plain hood without any mark. This was thought by Seba to be a male, but both male and female of different varieties are found marked alike.

No. 23. *Mogla Nagoo.*

Abdominal scuta 192, subcaudal squamæ 65 = 257.

This variety has received its name on account of frequenting the Caldiero hedges. The cervical scuta are spotted here and there with faint grayish spots, and the post-oculars are of a bluish-gray color.

No. 24. *Malle Nagoo.* (*Malle* is the Arabian name for jasmine.)

Abdominal scuta 191, subcaudal squamæ 62 = 253.

Color lighter brown than the preceding variety; scuta whiter and less spotted, but seven of the cervical scuta are entirely dark.

No. 25. *Camboo Nagoo.*

Abdominal scuta 186, subcaudal squamæ 60 = 246.

Some deviations observed in the shape of the laminæ; cervical scuta dark; trunk has a strong bluish cast.

No. 26. *Jonna Nagoo.* (*Jonna* is a small grain used as food for horses.)

Abdominal scuta 189, subcaudal squamæ, 57 = 246.

An orange color prevails in the skin of the hood; scuta of the neck are spotted with gray, six of the lower ones being wholly of a bluish-gray.

No. 27. *Nella Tas Pam.*

Abdominal scuta 186, subcaudal squamæ 62 = 248.

The black color on the hood is of an unusually deep shade,

and all the jugular scuta are of an unusually dark color, hence its name.

No. 28. *Kistna Nagoo.*

Abdominal scuta 186, subcaudal squamæ 63 = 249.

Of the three laminæ between the eyes the middle one is very broad, and the posterior pair (occipitals) are subovate. Five of the jugular scuta are dusky, and six of the cervical scuta are almost black.

No. 29. *Korie Nagoo.*

Abdominal scuta 184, subcaudal squamæ 57 = 241.

Supraciliaries remarkably narrow; the large occipitals oval; color of the trunk, more especially of the scuta, unusually bluish.

When the Cobra prepares itself for an attack it makes a full inhalation, by which, the whole body being inflated, the scales are separated from each other; when it exhales, the shrinkage of the body brings the scales together again, excepting upon the hood which remains permanently expanded. When erect and in readiness to strike, the head is at right angles to that part of the body below the hood, which in no way, however, protects or shields the former. The hood when expanded is convex on its posterior and concave on its anterior surface. This shape is retained with ease by the reptile, owing to the peculiar shape and arrangement of the vertebræ in this part of the spinal column, whose forked ends are interwoven in such a way as to give to the hood its concavo-convex form.

The Cobra seems to be the favorite species exhibited by the snakemen as a dancer. The showman, seated on the ground like a tailor on his bench, begins sounding his reed pipe, and at the same time takes the cover off the round, flat basket in which the snake is carried. If it is slow to come out, a tap on the head from the showman hastens its movements, otherwise it is unceremoniously emptied out of the basket. The piper commences a slow movement of the body from one side to the other, which the snake soon imitates, with its eyes fixed intently upon his right hand, which is shielded by the basket, and now and then pushed forward to provoke the Cobra to bite.

The movements of the reptile are graceful and amusing; but this exercise soon tires it, and a fresh one is at hand, in another basket, to take its place. These snakemen are always provided with their antidotes against bites; but when they allow themselves to be bitten, it is always by a snake whose fangs have been extracted, although bystanders are not supposed to be aware of this fact. In a wild state, music has little if any influence on the snake, it being trained, by a long and severe schooling,* for its subsequent exhibitions.

Several varieties of Cobra are known on the coast of Coromandel under different names. The species was supposed to exist in South America,† but it appears that it is found exclusively in India and Africa.

The Spectacle or Hood snake of North America has the skin about the neck loose, so that the head is completely enveloped by the fold of the skin when the former is drawn back. In this position the back of the hood shows a distinct "spectacle-mark." This variety is not venomous.

The following variety of Cobra occurs at Maunbhoom, India, called Airá Gahman (No. 30): Average length 4' 3½"; top of head purple brown, shading into a bright orange color on the lower half of hood; on the back are two faint shades of vinaceous brown, in stripes; spectacle-mark of a dark

---

* Kempfer, Amœnitates Exoticæ, p. 569.
† Gronovius, Zoophylacium, vol. i, p. 20.

brown, with white periphery; throat and underneath the spectacle-mark ashy brown; belly of a pinkish-white color.

Other varieties are as follows, viz.:

No. 31. *Manilag.* Average length 36½″. Not a common variety. Very slender neck and broad jaw; top of head light brown, shading into yellow in the hood and back; belly yellowish-white.

No. 32. *Bichá Jarmá Gahman.* Average length 3′ 11″. Head ruddy brown; hood reddish-brown; spectacle-mark yellow, with reddish-brown periphery.

No. 33. *Kaléy.* Average length 4′ 3″. Whole body black; has a ring on its hood; two bands on the throat and an entire collar below the hood vary in color in different individuals, from a creamy white to a dirty gray.

No. 34. *Káuta Káris Gahman.* Average length 4′. A very slim variety. Head vinaceous brown; yellow tinge in hood; spectacle-mark red; throat a purplish-brown.

No. 35. *Dudhiya Gahman.* Average length 44″. Top of head vinaceous brown, darker on the hood and lighter along the back; below, ashy white; spectacle-mark white, with a dark-brown well-defined periphery.

No. 36. *Parula.* A beautiful snake. Average length 3′ 5″. Head a light shade of purplish-brown; no spectacle-mark; hood a bright red.

No. 37. *Sarsa Gahman.* Common yellow Cobra; most abundant.

No. 38. *Charara Gahman.* A large yellow Cobra, but with very distinguishing marks, distinct from the preceding variety.

No. 39. *Basta Karicha Gahman.* Average length 4′ 2″. A very dark vinaceous brown; spectacle-mark white, bordered with light red.

The natives in India consider the Gokurrah, a snake which

frequents towns and cities, and the Keautiah, a snake of the fields or jungle, that the poison of the latter is thinner, and therefore kills quicker, and that of the former thicker, and though slower to take effect, none the less deadly. Both kinds incubate; as the snakemen affirm to digging them out of holes frequently, in which they are found covering their eggs.

Smith gives the following varieties of this genus occurring in South Africa, viz.:

No. 40. *Naja haje* (Spugh Schlange).
No. 41. " *lemniscatus.*
No. 42. " *naia angusticeps.*

No. 40, or the Spew snake, is said to make a prolonged blow and a hissing sound, thus ejecting its poison to a considerable distance. This is one of the most aggressive varieties of Ophidians.

One specimen in the collection of the Zoological Society of London is so absolutely untamable that, after two years' confinement in its cage, it flies at any one who approaches too near it.

NAJA OPHIOPHAGUS.

The *Naja ophiophagus* (Hamadryad) is the only species given, called Sunkerchor (or shell-breaker); in Orissa, Ai-ráz.

Naja bungarus, *Schlegel.*
" elaps, "
" vittata, *Elliott.*
Hamadryus ophiophagus, *Cantor.*
Trimeresurus ophiophagus, *Dum. and Bibr.*
Hamadryas elaps, *Günther.*

One of the largest and most formidable venomous snakes known.*

---

* Waterton, in his Wanderings, mentions a Vip. snake, named Lachmutus, found in British Guiana, and called the Bushmaster, or Curucucu,

The colors and variegations of this variety vary in different places.

The *first variety* is olive-green above; shields of the head, scales of the neck, and hinder parts of the body and tail edged with black; the trunk with numerous oblique alternate black and white bands, converging towards the head; lower parts marbled with blackish, or a uniform pale greenish color. Habitat, Bengal, Assam, the Malayan Peninsula, and Southern India.

Second variety is brownish-yellow, of a uniform shade in the fore-part; with scales black, edged posteriorly; each scale of the tail with a distinct white and black edged ocellus. Found only in the Philippine Islands and in Burmah.

Third variety is uniform brownish-black; scales of the hinder part of the body and tail somewhat lighter in the centre; chin and throat are yellow, but all the remaining lower parts are black. This variety is found in Borneo. Young snakes of these varieties may often be mistaken for other varieties.

The *H. naja ophiophagus* is most common in the damp climates of Assam, Bengal, Orissa, and Southern India; specimens have been found measuring from 8′ to 14′ in length. It is very fierce, and always ready not only to attack, but to pursue. The dilatable neck is not wholly peculiar to the Cobra, for this is also seen in the *Compsosoma radiatum*, and the *Tropidonatus macrophthalmus*, both innocuous Colubers.

N. elaps Bungarus.—In this species there are two varieties in India, viz.: the *B. Cœruleus* or *Krait*, or *Bungarum Pamah*, and the *B. fasciatus*. The Krait, probably, causes more

---

sometimes fourteen feet long, found on trees, very poisonous, and very beautifully marked. This is the Trigonocephalus Lach. of Dr. Hering.

deaths than any other kind, except the Cobra; this is the *Gedi Paragoodoo* of Russell. It is also called Dhomun Chiti, in Bengal, &c. :

    Pseudoboa cœrulea, *Schneider*.
    Boa Krait, *Williams*.
    " lineata, *Shaw*.
    Bung. cœruleus, *Dand.*
     " lividus, *Cantor*.
     " candidus, *Cantor*.
     " arcuatus, *Dum. and Bib.*
     " lineatus, *Günther*.

Günther describes three varieties, viz. :

First Variety.—Upper part of a uniform blackish-brown (B. lividus; Cantor). In young specimens the head is white, with a black line between the occipitals.

Second Variety.—A vertebral series of equidistant small white spots, from which narrow transverse streaks proceed; upper parts with narrow white streaks in pairs (B. arcuatus). The color of the dark parts varies from a steel-blue black to a chocolate brown; tongue white; iris black. A common species found in the whole of India. Fangs not so long as the Cobras, and its poison is less deadly; its length varies from 29" to 47". The Krait may be mistaken for the Lycodon Aulicus, an innocuous snake of similar appearance, but the former is always darker than the latter, and this lacks fangs.

N. Elaps Bung. fasc.—This is the Bungarum Pamah of Russell, or the Sankni or Rajsamp of some, also Koelia Krait.

    Bung. annularis, *Schleg., Daud.*
     " fasciatus, *Cantor*.
    Pseudoboa fasciata, *Schneider*.

This variety has been found 6 feet long. Body variegated by a series of black and steel-blue or light-gamboge yellow rings; lips and throat gamboge, tongue flesh color; fangs much smaller than the Cobra's, and its poison does not kill as quickly. Dogs bitten by this snake died in from 4 hours 28 minutes to 10 days.

N. E. Henurelaps bungaroides, Günther.—This is the only species known of the genus Henurelaps; length $15\frac{7}{8}''$; tail $1\frac{7}{8}''$; brought from the Khasya hills. This variety closely resembles the Bungarus.

Body subcylindrical, long, slender; head short, subtriangular, rounded rostrum; neck not dilatable; tail short; shields of the head normal, but the loreal is wanting; nostrils lateral, between two shields; eyes small, one pre-, two postoculars; scales smooth, slightly imbricate, and in 15 rows; vertebral series enlarged, hexagonal. Anal entire, subcaudal bifid; maxillaries with a grooved fang in front, and a small smooth tooth behind.

Jerdon says: "It has one white intercepted line, commencing on the vertical and extending to the throat on each side;" color of body a deep, rich madder-brown, and the bands yellow, growing more pale towards the anus.

N. Elaps. Callophis.—This species has several varieties, all venomous, very short fangs. Fowls bitten died in from 1 to 3 hours. The varieties are, viz.:

No. 1. E. Call. intestinalis. No. 4. E. Call. trimaculatus.
" 2. " Maclellandii. " 5. " nigrescens.
" 3. " annularis. " 6. " cerasinus.

They are all more or less distinguished by a bright color on the sombre hue of the general surface of the body. They are very sluggish, and not aggressive. Günther says they live principally upon the Calamaridæ (a harmless snake), which they resemble very much, viz.:

Body cylindrical; abdominal scuta 200 and more; head slightly depressed, not distinguishable from the neck; nostrils lateral, narrow, between two shields; eyes small, one pre-, two postoculars; temporals in a single longitudinal series; 6, 7, or 8 upper labials, the 3d and 4th entering the orbit; scales smooth, 13 rows, uniform, subcaudal, bifid; a grooved fang, without any teeth behind it.

The Callophides being, viz.:

Body cylindrical, of nearly the same uniform diameter, much elongated; abdominal scuta 200 and more; head slightly depressed, hardly distinguishable from neck; shields on head normal, occipitals somewhat elongate, loreal wanting; the single preocular forms a suture with the postnasal, extends on to the upper surface of the head, but does not reach the vertical. Two postoculars in contact with the anterior temporal. Upper labials eight or less, the third and fourth entering the orbits. Scales invariably in thirteen rows, smooth, polished, slightly imbricate; tail short, subcaudals bifid.

First Variety—N. E. CALL. INTESTINALIS. Found in Central India; habitat, Singapore. Elaps furcatus, *Schlegel, Schneider.*

Aspis intestinalis, *Laur., Syn., Amph.*
Maticora lineata, *Gray.*
Elaps intestinalis, *Cantor.*
Calloph. intestinalis, *Günther.*

And described by the latter as follows, viz.:

Head light-brown above, yellowish below, spotted with black on the sides; a vermilion black-edged band runs from the occiput to the tip of the tail; a buff-colored band with an upper and lower black border runs along the joining edges of the two outer series of scales; upper black border as broad as the stripe of reddish-gray ground color on the side of the back. *Belly* with alternate pale citron and black cross-bands; the darker shade extends over three or four scuta, the lighter over two. *Tail* sometimes marked with three black rings. Upper labials 6; abdominal scuta, 223–271; subcaudal squamæ, 24–26; length 2′; tail 1½″.

Dr. Stoliczka exhibited a specimen of this variety at a meeting of the Asiatic Society, in 1870, which had the poison-gland extended down one-third the length of the body.

Second Variety—N. E. CALL. MACLELLANDII.

Elaps Maclell., *Reinhard*, I. Nat. Hist., 1844.
"   personatus, Jour. As. Soc., 1855.
"   univergatus, *Günther*.
Call. univergatus, *Günther*, P. Zool. Soc., 1859.
"   Maclellandii.   "   "   " 1861.

Upper labials 7; temporals small $1 + 1 + 1$; anals bifid; *head* and neck black above, with a yellow cross-band behind eyes; *body* and tail reddish-brown, generally with a dark brown stripe down the entire length of the dorsum; *belly* yellowish with black cross-bands or quadrangular black spots. *Length* from 18″ to 26″. Three varieties are given by Günther, but they differ slightly in appearance.

Third Variety—N. E. CALL. ANNULARIS.

Length 19″, tail 2″; upper labials 6, temporals small $1 + 1 + 1$; first very narrow, third large; abdominal scuta 208, anal bifid; subcaudals 33. *Head* and neck black above, with a broad yellow cross-band behind the eyes; body and tail reddish-brown, with forty narrow equidistant black white-edged rings; these broadest at the tail, but at neck $\frac{1}{4}$ of an inch broad; on the belly they are the breadth of an abdominal scuta. *Belly* yellowish, with a black cross-band in the middle between the rings.

Fourth Variety—N. E. CALL. TRIMACULATUS.

Vipera trimac., *Dand.*, Rept.
Elaps   "   *Merr.*, Tent.
Coluber melanurus, *Shaw*.
Call. trim., *Günther*, P. Zool. Soc., 1859.

Length 12″, diameter $\frac{3}{8}$″; upper labials 6; temporals elongate $1 + 1$; abdominal scuta 258–274; anal bifid; subcaudal squamæ 35. *Color* light bay above; an indistinct line formed by minute brown dots along each series of scales. Upper side of head, neck, and a spot below each eye black; rostrum with some small irregular spots; a yellow spot on each temporal shield; a subtriangular yellow spot in the middle of the neck. The black of the neck edged with yellow behind; *tail* marbled with black below; has two black rings, each of which is variegated with yellow; *belly* red.

This is the Coluber, No. 8, of Russell.

Fifth Variety—E. CALL. NIGRESCENS.

E. Call. nigrescens, *Günther*.
" concinnus, *Beddome*.
" Malabariens, *Jerdon, Bedd*.

Vertical shields elongate; upper labials 7; temporals 1 + 1, the anterior of double size; abdominal scuta 232-247; subcaudal squamæ 33-42. Anal entire generally, but found bifid in one specimen. Length 4', tail 5''. Upper parts darkish ash or black; the lower part red, of a uniform shade; upper part of head symmetrically marbled with black; a black spot below the eye; another from the occipital to the angle of the mouth; a black horseshoe-like collar with its convexity forwards; a narrow black vertebral line slightly edged with yellow, runs from the collar to the tip of the tail; a series of small ovate black spots, indistinctly edged with whitish color, along each side of the trunk, disappearing towards anus; *tail* colored like body, without black rings.

Sixth Variety—E. CALL. CERASINUS. From Malabar; described by Major Beddome as follows:

Length $21\frac{1}{2}''$, tail 2''; abdominal scuta 228, subcaudal squamæ 13; rostral shield slightly produced back between the frontals; the latter touch the orbit; loreal and anteocular wanting; nostrils between two nasals; seven upper labials; 3d, 4th, and 5th very high; 3d and 4th enter the orbit; one small post-ocular vertical, hexagonal, elongated, pointed behind; supraciliary small; occipitals large, elongated, pointed behind, with a pair of large temporals on each side and entire; *back* purplish-brown with a pearl-like lustre, with forty transverse, broad, irregular-shaped black bands extending to the tail, equal distances apart, and though reaching down to the belly on the sides do not quite meet; the two lower rows of scales and belly of a bright cherry red; *head* black in front; *neck*, with the 5th, 6th, and 7th labials, and a portion of the occipitals, cherry red.

A North American variety of this genus is the *Naja Elaps Cuprocephalus* (nobis); Spanish, Cabeza de Cobre, Copperhead.

Agkistrodon Contortrix, B. & G.—This species is not known to have been found except in the Middle and Southwestern

States of North America, abounding particularly at or near the Bigbone Lick in Kentucky.

It is very much feared on account of the activity of its poison, and its bite is universally considered fatal.

Length generally 1 metre, but this varies from 75 to 125 centimetres. Head elliptical, ovoidal, short, somewhat flattened, black, and skin dark chocolate color, darker in the female than in the male. Belly of a coppery white; eyes small and reddish. The upper surface of the head is of the same color and lustre as a plate of polished copper tarnished. The dorsal region of the female is marked with longitudinal stripes, of a darker and a lighter shade alternated; one series of the stripes of the same color and shade as that of the male, and the other of a darker shade. These stripes are from 2 to 3 centimetres wide in their widest part, and diminish to ½ centimetre in width at the anus; towards the neck they disappear by graduation of shades into a dark brown band, which encircles the throat. The male is smaller and shorter, of a uniform color of the light stripes in the female, and for this reason has been considered as belonging to a different species.

This variety has fangs; the general disposition of the buccal parts being identical with those of the Crotalus.

This species is always found in marshy districts and near places frequented by wild animals for drinking. It almost invariably bites low, in contradistinction to the Crotalus, inflicting a wound in the region of the ankle-joint both in man and in animals. An ox bitten by this serpent makes two or more paces in advance, and then faces right about as if to make a reconnoissance of his enemy, who has in the meanwhile coiled his tail about the roots of a shrub or stump, and commenced a slow undulatory movement from side to side (like that described of the Rattlesnake), accompanied by a slow buzzing sound, similar to that produced by the locust; very faint at first, it grows louder by degrees, until in a few moments the victim begins the same motion of the head from side to side in unison with that of the reptile. The body soon partakes of the same movement; vertigo supervenes; the blood starts from the nostrils, and the animal falls to the ground; violent contractions of the muscles follow; the body begins to swell, and shortly death ensues.

Some people entertain the belief that this serpent is endowed with a power of enchantment, and this is not without a certain foundation, in the fact that all serpents are highly susceptible to magnetism and magnetic influences. In proof of this there are "Curers" who will calm and quiet the most uncontrollable fury of the most venomous snakes, by making a few "passes" with both hands, when the reptile approaches, coils himself up into a small coil, the head low, mouth closed, and thus remains quiet immediately under the "Curer's" hands.

One who has *seen this* must acknowledge that some like influence gave birth to the tales and legends of enchantment that are transmitted to us from former ages. One may safely venture to predict at no very distant day in the future, the application of magnetism for the cure of certain classes of diseases in conformity to the law of "similars." The instance recorded in Good's "Book of Nature," of the bird enticed into the jaws of the snake by its powers of fascination, has been repeated to me by different Curers; men who are unacquainted with each other, who hardly know how to read a little even in Spanish, much less in English, and whose sole knowledge of the habits of snakes and reptiles is derived from hours, months, and years of study of their peculiarities from nature itself, roaming about in the woods and forests.

It would exact no great effort to fill a volume with these "gleanings" treasured up in memory, but as I only propose to enumerate well-authenticated facts, and only such matters as have come under my immediate observation and experience, I forbear.

The species *N. Elaps Cuprocephalus* derives its name from

the color and lustre of the head bearing such a close resemblance to a surface of polished copper, as the name indicates.

N. ELAPS CORALLINUS, Vipera corallina. Spanish, Corál; French, Serpent coráil.

Coral Snake.

The Coral Snake is found in Mexico, Central America, and in all the lowlands of South America as far south as Southern Brazil, up to from 200 to 1000 metres above the level of the sea.

Of Ophidians it is the species possessed of the most brilliant colors, most harmoniously blended. By many persons it is considered to be exceedingly venomous; as many others, perhaps, do not believe that its bite produces fatal effects. This is undoubtedly owing to those climatic conditions which contribute in such a marked degree to modify the virulence of animal poisons in the same genera and species.

By reference to the tabular list of snakes in the different States of Colombia, it will be observed, where the climate varies from a torrid heat to the cold of a region of perpetual snow (14,000 to 18,000 feet above the level of the sea), *the same species occupy different relative positions in different places*, in accordance with the fact that the species are intended to occupy (in the table), each one its comparative position as regards the virulence of its poison, so far as it has been possible to institute a comparison between the different species.

Under these climatic conditions, the electro-magnetical states of the atmosphere exercise an important influence (as previously stated), modifying, in a remarkable degree, the effects of the poison in every species.

Thus, one may readily conclude that the same species may

be exceedingly venomous in *some* districts, and at certain periods, while in *other* sections, and under different conditions, it may be innocuous. Facts prove such a conclusion to be entirely legitimate.

The Coral Snake varies in *length* from 50 centimetres to 1 metre.

Body from 3 to 4 centimetres in diameter. *Head* somewhat broader than the body, somewhat flattened, and of the shape of two cones, joined at their bases with rounded apices; its superior surface covered by 11 shields of a whitish color, separated by black lines, and arranged in transverse series, as follows, viz.: (commencing at the nares) *two* small ones; *two* a little broader and wider; three still larger (in the edges and external periphery of the outer ones of which the eyes are placed); and, lastly, *four* still broader and wider. The shape, contour, number, and color of these scales is invariable in every individual of this species, male or female, pure or hybrid.

In the *female*, the posterior third of the head is marked with an entire band of small black scales, broader on its posterior edge, with a narrow white ring; this band is followed by one of coral-red, 7 centimetres wide, and extending from side to side of the belly; next to this is a black band, like the first, 2 centimetres wide, with a narrow white ring on each edge; this alternating one with another, there are from 12 to 16 bands terminating with the anus, the caudal part being black.

The *belly* is whitish-red along its central axis, and shaded out to meet the bands in graduations of their respective colors.

The *male* is shorter and of smaller diameter than the female; *head* smaller, more flattened, and without the posterior black ring but from the posterior third of its length to the caudal extremity; *body* of a uniform coral-red color; *belly* reddish-white in both male and female; it is covered with from 150 to 200 scuta, slightly curved towards the head; subcaudal squamæ, 20 to 24 pairs, divided through their centre by an irregular line.

Of more than fifty individuals of this variety examined in South America, the buccal parts are as follows, viz.: deprived of fangs; the triangular sac is wanting; one of the teeth (in the same relative position of the fang in the Rattlesnake, which is at the anterior third part of the longitudinal axis of the mouth), is slightly prolonged and curved backwards; this tooth has its anterior part deeply grooved from base to point; a minute filament terminating *in a sheath* (in line with this groove and at the base of the tooth), connects with a minute bladder (containing the poison), which lies in a concavity in the bony process of the upper jaw, and in the same relative position as in the Crotalus horridus. By a slight pressure on the point where the poison-bladder lies, the filament will be seen to protrude from its sheath, adjust itself to the groove down which it slides, and when it reaches the extremity of the tooth, a drop or more of poison flows from its point; at the same instant a small quantity of greenish, viscous mucus flows down towards the point of the tooth from a gland situated at its posterior part; a drop is seen to form on the point of the tooth; the reptile makes a hissing sound, with the mouth still open, and these drops of mucus are thus expelled.

This latter is bitter to the taste. In fact, as far as it is possible to judge (without entering into the details of a chemical analysis), it seems to be *gall mixed with saliva.* What are its use and functions? Evidently a small quantity of the poison *may possibly have remained* on the lower point of the tooth after its ejection by the filament, and this mucus is discharged by its apparatus upon just that portion of the tooth which *may have been moistened by the poison;* palpably and undoubtedly it is a disposition of nature to antidote the poison by a secretion generated in the reptile's own organs; this antidote is the gall, and experience proves it to be the most efficacious one known, as may be seen in another page. This conclusion was the result of a microscopical examination of a Copperhead, killed several years since near Covington, Ky.

But, according to Dr. Mure,* the variety existing in Brazil has the teeth alternated with small bifurcated valves, which cover the mouth of the secretory glands of the poison. It is to be hoped that some Brazilian naturalist will examine minutely *several individuals* of the species, to determine this point with precision, for its existence would furnish a characteristic which would place the variety found in the United States of Colombia and Northern South America under another name. Possibly, Dr. Mure mistook the salivary glands and duct for the secretory apparatus of the poison. The lower jaw is provided with two series of teeth, lower maxillary and lingual, which are furnished with short, fine, closely set teeth, all considerably curved backwards, the upper ones being the usual series of marginal maxillary and palatine teeth.

This species has no habits which lead it to choose one region or locality in preference to another, except that it does not frequent low, swampy places. Individuals are found in the walls of houses; under loose boards; in cribs of corn; in the thatch of the roof; under the pillows and clothes of the bed; in any kitchen utensil, and everywhere in the forest. The male is often seen copulating with the Viba. Aznfrada, and many other varieties; but the female is never found copulating with any but the male of her own species! This is a singular fact, and shows how nature preserves the kind by endowing the female with this instinctive preference.

Varieties of this species found in South Africa are, viz.:

No. 1. N. Elaps Sunderwallii.   1 sub-variety.
" 2.   "   hygeæ.   1   "
" 3.   "   dorsalis.   1   "
" 4.   "   aspidelaps rubricus. 1   "

---

* Pathogenesie Brésillienne.

The following occur in Australia, viz.:

No.				
1.	N. Elaps	diemenia	psammophis.	1 sub-variety.
2.	"	"	olivacea.	1 "
3.	"	"	reticulata.	1 "
4.	"	"	mulleri.	1 "
5.	"	"	superciliosa.	1 "
6.	"	"	torquata.	1 "
7.	"	pseudonaja nuchalis.		1 "
8.	"	furina callonotos.		
9.	"	" bimaculata.		
10.	"	brachyurophis australis.		

A variety occurs in Brazil, called:

11. N. Elaps lemiscatus (Labarri). Its skin is beautifully variegated with rainbow hues, which fade after death. This has fangs.

Nineteenth Genus—HYDROPHIDÆ.

Aquatic snakes, found in the China seas and in the Persian Gulf and coast of Madagascar. All have flat, two-edged, eel-shaped tails, and are exceedingly poisonous. Some varieties attain a length of 12 feet or more. Günther says they most abound in the seas between Southern China and Australia. They have generally small jaws, and smaller fangs than the land snakes. Cases of their bites have produced death in from 1 hour 15 minutes to 72 hours. The last was an unusual case, and the bitten person was administered alcohol freely during the time he remained alive, which undoubtedly prolonged greatly the action of the poison; at the same time, other causes (suggested on another page), may have combined to deprive it of its greatest activity at the time it was introduced into the blood.

The abdominal scuta in this species is generally not well marked; all have an eel-like appearance.

The *Platurus* seems to be a transitional link between aquatic and land snakes, and it has disproportionately large abdominal scuta. The eyes are exceedingly small, and have circular pupils, which latter covering the eye makes them almost blind when taken out of the water. The head shields are peculiar, and differ from those of land snakes. Nasals large; nostrils on the edge of rostrum; anterior frontals wanting; a single pair of frontals; one vertical; two supraciliaries; two occipitals; one ocular; one or two post-oculars; loreals none; labials irregularly arranged and divided.

This genus has seven species, which occur in India, &c., viz.:

No. 1.	Hydr.	platurus.	2 varieties.
2.	"	aipysurus.	4 "
3.	"	disteira.	2 "
4.	"	acalyptus.	5 "
5.	"	hydrophis.	33 "
6.	"	enhydrina.	2 "
7.	"	pelamis.	1 variety.
		Total,	49 varieties.

Nos. 1, 5, 6, and 7 occur in the Indian Ocean.

First Species—PLATURUS.

Günther gives two varieties:
No. 1. Hydrophis platurus scutatus.
2. " " Fischerii.

In this species the tail is not prehensile; body subcylindrical; scales imbricate, smooth, in from 19 to 25 series; subcaudal squamæ bifid; head shields regularly disposed; poison-fang short, and has sometimes a single large tooth behind it.

No. 1. *Hydrophis platurus scutatus.*

Coluber laticaudatus. *L. mus. Ad. Fried.*
Laticaudatus scutata. *Laur. syn. rept. (Cantor).*
Hydrus colubrinus. *Schneider.*
Platurus fasciatus. *Lat. rept.*
Hydrophis colubrinus. *Schlegel.*
Habitat, Bay of Bengal.

Length 5 ft. and upwards. Abdominal scuta 213 to 241; dorsal scales disposed in 21 to 23 longitudinal rows; body surrounded by from 25 to 50 black rings; an azygos shield between the posterior frontals; crown of head black; first and second black mark on head and neck are joined below by a black longitudinal band commencing from the chin; rostrum and side of the head yellow, with a black band running across the eye.

No. 2. *Hydrophis platurus Fischerii,* Jan.

Habitat, the tidal streams near Calcutta.

Abdominal scuta 232 to 241; scales in 19 rows; trunk marked by 33 to 36 black rings, which are broader than the interstices; a black band crosses the occiput and extends forward over the vertical plate and lower jaw, but is not confluent with the adjacent ring; upper part of rostrum yellow; upper labials black.

Two varieties of this species occur in Australia.
Further species occurring in the seas about Australia are:

No. 2. Aipysurus.   3 varieties.
3. Disteira dol.   1 variety.
4. Acolyptus.   1   "
5. Hydrophis.   4 varieties.
6. Enhydrina.   1 variety.
7. Pelamis bicolor.   1   "

And Krefft adds—

No. 8. Emydocephalus. 2 varieties.

Making in all 15 varieties already classified.

HYDROPHIDÆ.

Fifth Species—HYDR. HYDROPHIS, *Daud.*

This species has the following varieties, viz.:

No.			
1.	H. hydr.	Jerdonii.	Indian coast and Penang.
2.	"	Stokesii.	Doubtful.
3.	"	major.	Indian Ocean.
4.	"	robustia.	Indian Ocean.
5.	"	cœrulescens.	Bay of Bengal; Penang.
6.	"	aspera.	Singapore.
7.	"	spiralis.	Indian Ocean.
8.	"	cyanocincta.	Indian Ocean, Ceylon, and Bay of Bengal.
9.	"	subcincta.	Indian Ocean.
10.	"	nigrocincta.	Bay of Bengal.
11.	"	torquata.	Bay of Bengal, Penang, Soonderbuns, Indian Ocean.
12.	"	chloris.	Soonderbuns, Bay of Bengal.
13.	"	Lindsayii.	Indian coast.
14.	"	latifasciata.	Mergui.
15.	"	coronata.	Bay of Bengal, Soonderbuns.
16.	"	diadema.	Indian Ocean.
17.	"	gracilis.	Indian coast, Bay of Bengal.
18.	"	fasciata.	Indian coast, Bay of Bengal.
19.	"	Cantoris.	Penang.
20.	"	lapemoides.	Indian coast; Ceylon.
21.	"	longiceps.	Indian Ocean.
22.	"	stricticollis.	Doubtful.
23.	"	ornata.	Indian Ocean.

No. 24. H. hydr. Elliottii.   Indian coast; Ceylon.
25.   "   pachycereus.   Indian Ocean.
26.   "   viperina.   Indian coast.
27.   "   curta.   Indian coast, near Pooree.
28.   "   Hardwickii.   Penang.
29.   "   Fayreriana.*   Indian coast, Bay of Bengal, Pooree.
30.   "   tuberculata.*   Tidal streams, near Calcutta.
31.   "   crassicolis.*   Bay of Bengal.
32.   "   Stewartii.*   Bay of Bengal.
33.   "   nigra.*   Bay of Bengal.

Nos. 1, 4, 8, 10, 12, 15, 22, 27 are Günther's, and Nos. 29, 30, 31, 32, and 33, are classified by Dr. Anderson.

Experiments made with the poison of several of these varieties prove it to be very deadly. Fangs are small and grooved, the involution not being so complete as in the fangs of land snakes. They possess a strong family resemblance to each other. Günther's general description of the genus is as follows, viz.:

Posterior part of body strongly compressed; head short, or of a moderate length, shielded above; one pair of frontals; nostrils superior in a single nasal; both contiguous; scales imbricate or not so, not polished, generally carinated; ventral shields very narrow, or quite rudimentary, or absent; lower jaw without the notch in front.

Günther describes the following, viz.:

Species No. 1. *Hydroph. Jerdonii.*

Hydrus nigrocinctus, *Cantor.*
Kerilia Jerdonii, *Gray.*

Abdominal scuta 235–238, bituberculate. Head short; rostrum de-

---

* Anderson.

clivous, rather pointed. Body covered with 19-21 rows of large, highly carinated, imbricated scales, higher than long, with apex slightly truncated; frontal shields small, not much larger than the preocular; one postocular; five upper labials, the third and fourth of which enter the orbit, the last one below the postocular; two or three large temporals on the side of each occipital, the anterior of which enters the labial margin behind the fifth labial; anal scuta small; terminal scale of the tail large; a series of seven simple teeth behind the fang in front; trunk variegated with 34 to 36 black cross-bands, broadest on the back, and extending to the belly in young individuals.

The genus Hydrophis is sufficiently described for the purpose by what has already been given. The description of each species can only be carried out in detail in a zoological treatise.

Sixth Species—ENHYDRINA, *Gray*.

This genus differs from the Hydrophis only in the deep cleft in the lower jaw where the mandibles do not unite, and the fold of the integument connecting them forms a deep notch.

Dr. Russell describes two species as Valakadyen and Hooghly Pattee:

 Enhydrina Bengalensis, *Gray*.
   " valakadyen, "
 Valakadyen, *Russell*.
 Hydrus valakadyen, *Boie*.
 Hydrophis schistosa, *Schleg., Fisher, Dum. and Bib.*
 Hydrus schistosa, *Cantor*.
 Hydroph. Bengalensis, *Gray*.
   " subfasciata, "
 Thalassophis Werrerii, *Schmidt*.

Dr. Russell says these have no fangs, but the specimens examined show a short fang with the poison-channel open throughout a part of its length. (For description, see Russell, vol. ii, 10 and 11.)

Pelamis.—The most common species of sea-snakes. Habitat, the Bay of Bengal, where it is called Kullundur. The only variety given is—

>Pelamis platurus, *Linn.*
>" bicolor, *Daud, Günther.*
>Hydrus bicolor, *Schneider.*
>Hydrophis variegata, *Schlegel.*
>" pelamis, "
>Pelamis ornata, *Gray.*

Habitat, Pooree.

Length 12½". Head flat, with a long spatula-shaped rostrum. Neck stout; nasal shields contiguous, pierced by the nostrils in their posterior edge; one pair of frontals; abdominal scuta 378 to 440; scales in 45 to 51 series. Color variable.

Günther describes four varieties:

First Variety.—*Color* black above, sides and belly uniform brownish olive; tail with black spots. *Second Variety.*—*Back* black, belly and sides brown, separated by a black and yellow band, large spots posteriorly. *Third Variety.*—The black color on the back becomes narrow and sinuous behind the middle of the body; posteriorly a dorsal series of rhombic confluent spots, sides and belly with an irregular series of rounded black or brown spots. *Fourth Variety.*—Yellow, with about fifty brown, black-edged cross-bands extending to the belly, which is crossed by narrow, vertical, brownish-black streaks, alternating with the dorsal bands, some of which are confluent, forming a zigzag band; head yellow variegated with black.

OPHIDIA VIPERIFORMES.

This suborder has two genera, viz.:
>No. 1. Crotalidæ, and No. 2. Viperidæ.

CROTALIDÆ.

TWENTIETH GENUS—CROTALIDÆ (*Pit Vipers*).

It is made up of the following species, viz.:

		Varieties.	
	1. Trimeresurus,	9	
	2. Peltopelor,	1	India.
	3. Halys,	2	
	4. Hypnale,	1	
	5. Crotalus,	3	United States.
		9	"
Crotalidæ.		1	Mexico.
(11 species	6. Caudisona,	1	New Mexico.
and 39 va-		1	Venezuela.
rieties.)		1	Guiana and Brazil.
		1	" " Mexico.
	7. Craspedocephalus,	3	2 Brazil, 1 Martinique.
	8. Fasciatus,	3	U. S. of Colombia.
	9. Crotaphopheltus rufescens,	1	South Africa.
	10. Callosclasma rhodostoma,	1	Java and Siam.
	11. Crotalophorus miliaris,	1	Asia (Kuhl).
	Total,	39	varieties.

Of this genus the only species found in India that has the vestige of a caudal appendage is the Halys; and in this it is reduced to a simple horny spine or point.

First Species—CROTALUS TRIMERESURUS, is made up of—

No. 1. Crotalus Trimeresurus gramineus, *Günther*.
" erythrurus, "
" carinatus, "
" anamallensis, "
" monticola, "
" strigatus, "
" mucrosquamatus, "
" Andersonii, "

Crotalus Trimeresurus.—Although the individuals of this species are fierce and venomous, yet few deaths are ascribed to their bites. They are probably ovoviviparous. Dr. Günther thinks the degree of danger in case of a bite depends upon the size and fierceness of the snake. Facts do not seem to warrant this conclusion, but the experiments with poisons point decidedly to the Cobra poison as the most deadly, and as being unique in its action; but there are well-authenticated cases of the South American Echis striata producing death in less than three minutes, and this is a snake not more than 17″ to 20″ long, and ½″ in diameter, but with disproportionately large fangs. One thing that is peculiar to this species is that they adapt their color to the locality in which they live, like the well-known tree toad.

Crotalus Trimeresurus gramineus.—This is described as the Boodroo Pam by Dr. Russell, under two varieties.

    Vipera viridis, *Daud, Rept.*
      " gramin., *Cantor.*
    Trimeres. viridis, *Gray.*
      " elegans, "
    Coluber gramineus, *Shaw.*
    Trimeres. gram., *Günther.*

Common in Assam and the Khasya hills.

Length from 18″ to 32″; tail 2⅜″; abdominal scuta 158 to 170; subcaudal squamæ 58 to 71. Scales on the head are small and smooth, or indistinctly keeled. *Color* of a black grass-green, lighter on sides, and greenish-white on belly; tail is sometimes red; a yellowish or brick-red line runs along this outer series of scales. It is smaller than the *Trimeres. carinatus.*

    Crotalus Trimeres. erythrurus.

    Trigonocephalus erythrurus, *Cantor.*
    Trimeres. albolabris, *Gray.*
    Trigonocephalus viridis, *Schleg.*

This is the species described by Dr. Russell (in vol. ii, pl. 20) as Boodroo Pam.

Head elongate-oval, much depressed; *tail* and rostrum white; the lateral line white bordered with purple or greenish below. *Color* grass green, lighter on sides and belly; scales on body highly carinated, in 21 to 23 series. Abounds in the delta of the Ganges, Moulmein, Penang, and Java, and probably on the Nicobar Islands.

Crotalus Trimeres. carinatus.

Trimeres. carinatus, *Gray.*
    "   bicolor,   "
    "   porphyraceus, *Blyth.*

This species is found in Bengal, Sikkim, Himalaya, and Burmah, resembling closely the Trimeres. gramineus.

Length 36″, diameter 1⅝″. Scales on body in 25 rows prominently carinated. *Color* dark grass green, darker on head, and tail lighter green below, approaching to white on belly. *Head* broad, triangular, covered with small carinated scales; the second upper labials form the anterior margin of the loreal pit. There is one well-developed azygos shield between the supranasals. Abdominal scuta 164 to 169; subcaudal squamæ 54 to 60.

The Trimeres. carinatus, Trimeres. bicolor, and Trimeres. porphyraceus, seem to have slight differences in appearance, &c., although classed as distinct species.

Crotalus Trimeres. anamallensis, generally 24″ long or less.

Abdominal scuta 148 to 158; subcaudal squamæ 51 to 55. *Color* generally yellowish-green, with a dorsal series of large rhombic black spots, each spot subdivided by or variegated with yellow. Upper side of *head* marbled with black or a uniform green; a black or brown band runs from the back edge of the eye to the corner of the mouth; supraciliary with one or two black cross-streaks; belly yellowish-green with numerous yellow and black spots along its side; tail black, with yellow and green spots.

Crotalus Trimeres. monticola.
Parias maculata, Gray.

Length 23″ to 29″; diameter 1″; abdominal scuta 137 to 141; subcaudal squamæ 41, and with 23 series of black scales. Those on head smooth, on body slightly carinated; supraciliaries very large. *Color* in some individuals pale brown, with a vertebral row of large, square, dark brown maculæ; along the sides a row of small dark spots; a pale streak across the temples. *Belly* dark mottled. Other specimens are of a cinnamon gray, beautifully marked with dark square spots, with the common U-shaped mark on the neck of a whitish color. Eyes small; supranasals separated by two shields. Specimens were from Darjeeling.

Crotalus Trimeres. strigatus, Gray.
Trigonocephalus Nilgherriensis, Jerdon.

This species is found on the Nilgherries and Deccan, and also in Ootacamund.

Length not exceeding 19″. Abdominal scuta 136 to 142; subcaudal squamæ 31 to 46. Whitish, horseshoe-shaped mark on the neck; brown with dark markings, a dark brown band leading from eye to neck; whole upper surface of head covered with small nearly smooth scales; nine or ten upper labials; scales carinate in 21 series. *Tail* terminates in a conical euplike scale.

Crotalus Trimeres. mucrosquamatus.
Trigonocephalus mucrosquamatus, Cantor.

Habitat, Caga hills and Assam. Length not given. Cantor's description is, viz.:

Abdominal scuta 218; subcaudal squamæ 91. *Color* brownish-gray above, with black, white-edged rings, covered with oval, slightly carinated, pointed, imbricate scales, whitish beneath, dotted with black.

Crotalus Trimeres. Andersonii, Theobald.

Habitat, Assam.

Length 20″; *tail* 2¾″. Abdominal scuta 182; subcaudal squamæ 56 to

71. *Scales* carinated on body, 25 rows; second upper labial forms anterior margin of the preorbital pit; supranasals separated by an azygos shield. *Color* above and below a rich brown. *Belly* and sides conspicuously marked with white spots.

Crotalus Trimeres. Wardii.

Jerdon proposes this name for a species described as—

Greenish with purplish-brown diamond spots on back and sides. Length 11″ to 14″.

Habitat, Nilgherries.

Second Species—CROTALUS PELTOPELOR (*Günther*).

Crotalus Peltopelor macrolephis.
" Trimeres. " Beddome.

Habitat, Anamallay Mountains, at from 4000 to 6000 feet elevation.

Length 21″; *tail* 4¾″. Abdominal scuta 133 to 138; subcaudal squamæ 53 to 56. *Rostral* triangular, erect. *Head* covered with large plate-like scales. *Color* dark green, lighter below; last row of scales on each side white from neck to tail; scales in 12 to 14 rows, smallest ones form the white stripe; all pointed and highly carinated.

Third Species—CROTALUS HALYS (*Gray*).

Habitat, Hindostan.

This species is distinguished by a broad, obtuse *head* covered with shields in 23 to 27 series. Carinated scales; has squamæ. *Tail* short, terminates in a spine or thornlike point.

This species has the following varieties, viz.:

No. 1. Crotalus Halys Elliottii.
2. " " Himalayanus.

No. 1. *Crotalus Halys Elliottii.*

Trigonocephalus Elliottii (*Jerdon*), is:

Length two feet and upwards. Abdominal scuta 151; subcaudal

squamæ 43; scales in 23 rows. *Color* olive-green above, pearl-white on the belly.

Habitat, Nilgherries.

No. 2. *Crotalus Halys Himalayanus.*

Trigonocephalus affinis (Günther). Habitat, the Northwest Himalayas.

Length 23 to 34, diameter ¾″; abdominal scuta 162 to 166, subcaudal squamæ 43 to 51. Rostrum broader than long; nose protruding; rostral oblique, higher than broad; frontals well developed, entire; anterior frontals short; transverse tapering on the sides, together form a crescent shape; posterior frontals large, pointed in front, rounded behind; occipitals small, rounded. Five upper labials, a sixth and seventh confluent with the temporals; the second small, not entering the margin of the facial pit; a third enters the orbit; three large temporals, the two hinder ones forming a portion of the lip; body round; tail terminates in a long spine. *Color* dark brown, with large band-like spots along the back, scarcely distinguishable except by their black edges. *Belly* almost entirely black marbled with yellow; a broad blackish band runs from the eye along the temporal shields to the angle of the mouth, with a narrow black and white edge above and below; lower labials marked with dun yellow and black.

This species is found, according to Stoliczka, at an elevation of 10,000 feet.

Fourth Species—CROTALUS HYPNALE (*Fitz*).

Variety *Crotalus Hypnale nepa*, called Carawilla in Southern India, or Coluber nepa, *Laur.**

        Carawala, *Davy.*
        Cophias hypnale, *Merr. tent.*
        Trigonocephalus hypnale, *Wagl.*
        Trimaculatus (?) Ceylonensis, *Gray.*
        Trigonocephalus zara, *Gray.*

Günther's description is, viz.: Habitat, Southern India, Malabar, Anamallay Mountains, and Ceylon.

---

* Described also by Russell, vol. ii, pl. 22.

Length 19″, tail 2½″; abdominal scuta 140 to 152, subcaudal squamæ 31 to 45. *Color* brown, gray, or reddish olive, with a double dorsal series of brown or black spots; those of both sides sometimes confluent with cross-bands. Sides and belly marbled and dotted with brown or black; upper lip brown or black, well marked by a darker line running from behind the eye to angle of the mouth; a whitish temporal streak above the dark line, sometimes continued along the side of the neck with an interrupted brown band above and below it. *Chin* and *throat* blackish or brownish, variegated with yellow or gray.

CROTALUS FASCIATUS.

The Banded Rattlesnake, found in the United States of Colombia. The varieties which abound in the Northern and Northwestern States of America are the most venomous; their poison is much more active than those found in tropical climates. The former are found of from two to three and one-half metres in length. (One was killed near Venice, Erie County, Ohio, in 1850, which measured 12 feet 4 inches extreme length.)

It abounds more particularly in Western New York, Western Pennsylvania, Ohio, Michigan, Indiana, Illinois, Wisconsin, Iowa, Missouri, Arkansas, Mississippi, and in Colorado, Nevada, Oregon, California, and the base of the Rocky Mountains, on both the eastern and western slopes. Following down the chain of mountains from California, through Mexico, Central America, and the Isthmus of Darien to the Southwestern Andes, this species still abounds at from 300 to 1000 or 1500 metres above the level of the sea; from Darien to the mouth of the Amazon, and from the plains of Casanare to the southern boundary of Brazil.

Its usual length is 150 centimetres (5 feet nearly). *Head* ovaloid, or triangular, partly covered with small tile-shaped scales. *Eyes* large in proportion to the size of the head; dorsal region of a reddish-brown, shaded with yellow sides. *Belly* of a much lighter shade, which changes to a lead-colored white below. Sides and back covered with from 30 to

36 rows of rhomboidal scales of a nearly uniform shape and size; dorsal scales slightly oval, lanceolated; caudal scales rhomboidal, but smaller.

The *belly* is covered with from 130 to 212 consecutive half rings or semi-elliptical scales (scuta) extending from the throat to the anus. *Body* thick, of a conical shape (in section), diminishing in size from its centre to the tail, and slightly reduced in diameter at the neck.

The caudal extremity is provided with a series of intersected hollow rings or capsules, so arranged that they have great freedom of motion and flexibility (these rings are composed of a substance similar in nature to horn); a new one of which is added every year, so that the number of composing rings of the rattle (cascabel) indicates the snake's age.

I have seen individuals with from 5 to 21 capsules, and others are found at times which have from 25 to 40, and that individuals of this genus do attain an age of more than forty years seems to be a fact established beyond all reasonable doubt.

The lower teeth are disposed in two concentric semi-ellipses (maxillary and lingual series); the upper ones in two series (maxillary and palatal); when the mouth is closed those in the lower series lie between those of the upper series in the spaces just outside and just inside them; all the teeth are curved backwards, longer or shorter according to the size of the individual, but of nearly an equal length. From each side of the upper jaw, and just back of a vertical line with the eye, there is a triangular-shaped bone to which the fang is attached; this is curved backwards, and when erect its chord forms nearly a right angle with the roof of the mouth or axial line of the upper jaw.

On the posterior side of its base is an articulation by which it is attached to the forward end of the external pterygoid bone; this articulation is a fixed point upon which the fang hinges, its erection and depression being effected by the corresponding muscles attached to the sides of the angle of the mouth, and controlled by the will of the reptile.

This fang is (in large individuals) from five-eighths of an inch to an inch in length, with a broad base or root, and very securely attached to one of the side-bones of the head. These bones are attached to the anterior and posterior shield-like bones of the head by a strong but flexible cartilage, which, yielding in every direction, permits a certain amount of movement to the base of the fang. The external portion of the fang is exceedingly hard and tough in its anterior part; and close to its point of attachment to the side-bones, it is perforated by a minute canal or tube, which extends down to the extreme lower point of the fang, terminating in an opening or slit of a double convex shape. This tube is entirely disconnected from all the inner portion of the root, nerves, or bloodvessels which nourish its bony mass. Just back of and below the eye, there is a slight concavity or depression in the upper mandibule, which contains a small sac or diminutive bladder, containing from six to twenty-five drops of poison. This small bladder terminates on one side in a threadlike tube, which passes through a perforation in the posterior wall of its concavity, and, passing inside and behind the buccal glands, is attached to the upper mouth of the channel in the fang.

On the other side the bladder is connected with the poison-gland and its secretory apparatus, which is so clearly described by Dr. Mitchell.* The South American varieties have a distinctly formed poison sac or bladder, as just described; and I have frequently taken from a snake six feet long twenty drops of venom from each fang, making *forty drops in all*. Although Dr. Mitchell says he could never procure more than five or six drops at a time, and then only

---

* Researches, &c , on the Venom of the Rattlesnake.

at the will of the snake; I can say that I never found any difficulty in extracting from any snake with fangs all the poison held in deposit in the poison-sac by pressure upon the external parts, and that in so doing I have never been able to discover that any act of volition on the part of the reptile could prevent it.

A *perfectly healthy person*, or one whose digestive organs are *perfectly sound*, can take two or three drops of this poison in a teaspoonful of water, *without its producing any fatal effects;* but should any organic lesion of any of the parts of the digestive apparatus exist, it would then act toxically, by being absorbed into the part where the lesion exists, and prove fatal in precisely the same manner as by the bite of the snake.

The buccal glands, which are attached to the periphery of the upper mandible on each side, by contraction press upon the poison-bladder, and thus produce a flow of the liquid through the extreme point of the fang, introducing it, by this arrangement, into the *deepest* part of the wound made by the latter.

A fold of membranous texture is attached from the anterior point of the attachment of the buccal gland to the posterior angle of the mouth, on the *inside* of the fang, and also along the periphery of the upper part of the mouth and *outside* of the fang, this forming a long triangular-shaped sac, in which the fang is inclosed when distended, which in its lower apex is perforated by a small hole surrounded by an elastic ring or sphincter of a fibrous nature. When this is expanded and the sac contracted, the fang protrudes; but when the sac is expanded, the contraction of the fibrous ring closes the little opening completely. The posterior part of the triangular sac has its two opposite sides united by a transverse membrane, which runs *nearly up to the point of attachment of the two sides*, thus dividing the sac into an anterior and posterior part.

A fibrous filament has one of its ends attached to the posterior extremity of the triangular sac and the anterior extremity attached to the base of the fang; along this filament are suspended a series of fangs, six or seven in number, the one next the fixed one being somewhat shorter, and the successive ones diminishing in size, the last being scarcely a line (one-eighth of an inch) in length.

By this beautiful and curious disposition of nature, the serpent can use its teeth and mouth for mastication, without any particle of food coming in contact with the apparatus which contains and ejects the poison. If a fang is broken or falls off, the filament contracts and draws a spare one to its place, which soon becomes fixed, and the apparatus is again in working order.

The color and variegations of the skin of this species differ in different places, those found in North America (as previously stated) having yellowish sides, covered with chocolate-colored spots, and a reddish-brown back. The varieties which are found in the United States of Colombia (*Crot. fasciatus*) are of a dun or seedy-black color on the back, and the whole length of the body divided into alternate stripes or bands, about three inches in width, beginning close to the *head:* the first one is of a smooth green; the next of an identical color with the back; the following one of the same color as the first band; and so on to the anus,—the last band being greenish and the rattle a dun color. The belly is greenish-white on the sides, and this shades down imperceptibly to a pale lead-color below.

These South American varieties are shorter than those found in the temperate zones, and are always found on hillsides where the surface is broken, and on stony or rocky ground. During the rainy season their poison is not considered so virulent in its action as during the dry season; but the female when pregnant is much more venomous than during any other period. The most expert "Curers" say that *their bite is not venomous when the atmosphere is surcharged with electricity;* and as during the wet season thunder and lightning almost invariably accompany every rainstorm, this may account for the fact previously stated.

In this connection, the question naturally suggests itself, will not electrical shocks destroy the effects of the poison? Experiments to determine this point would be interesting, and might develop some new fact about the action of the poison.

The genus *Crotalus* derives its name from a strong fetid smell, like musk, which the reptile emits when provoked: this smell is supposed or said to produce a stupefying effect. It never bites without giving previous warning by a shake of the rattle, and blows of the axe in felling trees bring one or more (that chance to be in the vicinity) to the spot. It selects a branch or shrub around which to coil the tail; makes a peculiar hissing noise with the mouth, and instantly "springs its rattle," measures its distance, and, with jaws distended to their utmost, its forked tongue bobbing in and out, like the needle of a sewing machine in motion, and eyes that seem to flash out sparks of fire, it sways the head and body from side to side with a slow undulatory motion that fascinates the sight, and then makes its final spring upon its prey. Its movements are comparatively slow, and the swaying motions of the head from side to side are, *per se,* graceful and even majestic.

Any person threatened by an attack of the Rattlesnake, however, can turn the reptile aside and avoid the danger by simply throwing a hat down in the snake's path; it will immediately coil itself around it, and wait for you to try and take it away. It is a common custom for the Curers, when they meet with one in the woods, to throw down a hat, and then go to the house or village to get some one to kill the snake for them, as they are superstitious about killing one *themselves;* they say it makes them fail oftener in performing cures!

The Crotalus "bites high," to use the Curers' expression. It is said always to wound *above the knee*, and by judging of *the distance of the wound from the ground*, they calculate whether the individual was large, medium-sized, or small. This rule appears to be deduced entirely from observation and experience, and it is not strange that these Curers are almost worshipped by the common people (like the magicians and soothsayers of olden times), when you see them examine a person bitten, and hear them say: "The snake will be found in such a place; it is so long" (measuring the length on the floor), and learn that the reptile has been found at or near the spot indicated, and is—with little difference—of the length foretold.

The poison of the Rattlesnake has been considered to be the specific cure for leprosy; but an experiment, made a few years since in Brazil by some allopathic physicians,* upon one M. J. Machado, added another name to the list of victims sacrificed under the protection of the law, in the name of science, without the results of the trial or experiment having given to the world aught in recompense for such heroic abnegation.

The *Crotalus Craspedocephalus Brasiliensis*, found in Brazil, is said to be very deadly; but no description of it has, as yet, been given.

---

* See Mure's Pathogenesie Brésillienne.

The *Crotalus Craspedocephalus lanceolatus*, or Yellow snake of Martinique, called Fer de lance, or Lance-headed viper, has been long considered in Europe to be the *Trigonocephala lachesis* of Dr. Constantine Hering; this is an error. For a description of the *Vip. T. Lach.*, see page 94.

Günther says that the varieties of this genus found in India have the extremity of the tail armed with a ball, or onion-shaped appendage, and not with a regular series of capsules, like the *Crotalus horridus* and *Crotalus fasciatus*. A like appendage occurs in some individuals of the latter species, very rarely, however.

The genus Crotalus is called a classification of Pit Vipers, because the varieties choose their dwelling-place in pits and caves, most generally frequented by burrowing or small animals; as, for example, occurs on the Plains of the Great West of North America, where it is a well-known fact that the caves of the prairie-dogs are always tenanted by a dog, a rabbit, and a rattlesnake.

Twenty-first Genus—VIPERIDÆ.

This genus has a great number of species already classified, and there are many varieties, particularly in South America, which have as yet received no scientific name. These are the true vipers, all provided with a poison of a nature more or less deadly; some have fangs, others are not provided with them, but have their buccal parts and poison-apparatus similarly disposed to the *N. E. Cuprocephalus*. The head is generally broad, triangular, and covered with small scales rather oftener than with shields; body short, rather thick in some varieties; in others long, slim, and of graceful proportions. A deep pit in the loreal region, which characterizes the Crotalidæ, is here wanting. It is very difficult to say, positively, which of the vipers has the most deadly poison, for, in other places, reference is made to cases where one and the other species have killed men in from *three to five minutes*, which facts rank them in deadliness beside the dreaded *Naja tripudians;* but these cases must be judged in the light of the fact that, when introduced into one of the great arteries, or rather, veins, any poison (having its deadly principle fully developed) kills with great quickness, hardly giving time to apply or administer any remedy with a hope of saving life.

The following are the species, &c., of this genus, viz.:

No.				Varieties.
1. Vipera Daboia Russellii (Günth.),				1. India.
2. Vipera Echis,	Carinata (Günth.), Striata (nobis), Variegata,	9	2. India and S. America. 4. South America. 3. " "
3. Vipera Echidna,	Inornata, Arietans, Atropos, Atropoides,	4	1. " " 1. " " 1. " " 1. " "
4. Vipera Pseudechis,	Porphyriacus, Australis, Scutellatus, Carinata, Major,	6	1. Australia. 1. " 1. " 1. South America. 2. " "
5. Vipera Tropidechis carinata,			1. Australia.
6. Vipera Dendraspis angusticeps (Wood).			
7. Vipera Actractaspis irregularis	"		South Africa.
8. Vipera heterodon,	Niger, Platyrhinus,	3	2. North America. 1. " "
9. Vipera achetulla laiocercus (Borneo), or Ahætulla.			1.
10. Vipera bucephalus capensis (Boomschlange)			1. South Africa.
11. Vipera Lachesis,	Bufocephalus, Trigonocephalus, Niger, Acuaticus, Variegata, Dryiophis, Os flavus,	7	1. South America. 1. " " 1. " " 1. " " 1. " " 1. " " 1. " "

OPHIDIANS.

No.				Varieties.	
12. Vipera Calamaris,	{ Venenosus, Ven. rubrum,	2	{ 1. 1.	South America. " "
13. Vipera Acuaticus colubriformis carinata.				" "	
14. Vipera brachysoma,	{ Diadema, Triste,	2	{ 1. 1.	Australia. "	
15. Vipera hoplocephalus, ...			18.	"	
16. Vipera petrodymon cucultatum, ...			1.	"	
17. Vipera Cachophis,	{ Krefftii, Fordii, Harriettæ, Blackmanii,	4	{ 1. 1. 1. 1.	" " " "	
18. Vipera Clotho,	{ Arietans, Cornuta,	2	{ 1. 1.	South Africa. " "
19. Vipera Cerastes caudalis, ...			1.	" "	
20. Vipera Cenchris piscivorus, ...			1.	United States of Am.	
21. Vipera Aspis (Aspor chersea), ...			1.	Europe.	
22. Vipera pelias berus (Wood), ...				"	
23. Vipera acanthophis antarctica (Wood), .			1.	Australia.	
24. Vipera lnpophrys, ...			1.	"	
25. Vipera ammodytes (Sandnatter), ...			1.	India.	
Total, ...			63 varieties.		

First Species—VIPERA DABOIA RUSSELLII; the only variety known, called Tic-polonga (*Davy*).

Daboia Russellii, or Tic-polonga, *Günther.*
Coluber Russellii, *Shaw.*
Vipera elegans, *Daud.*
"   daboia,   "
Daboia elegans, *Gray.*
"   pulchella,   "
"   Russellii,   "

Abdominal scuta 163 to 170, subcaudal squamæ 45 to 60 = 208 to 230. Length from 3′ 4″ to 4′ 2″.

This is also called Uloo Bora, Jesser, Siah Chunder, Amaiter, and known to Europeans in India as Cobra Monil. It is described by Russell as the Katuka Rekula Poda.

Dr. Fayrer kept one in a box for a year without food or water. Dr. Russell's description is as follows:

Abdominal scuta 168, subcaudal squamae 59 = 227. Length 4′ 2″. Tail tapers to a point. *Head* much broader than neck, gibbous or swelling behind, depressed above, compressed at sides, and narrowing from the eyes terminates in an obtuse rostrum faced with a pyramidal emarginate rostral; the labial and subjugular squamae are large and smooth, but the rest of the head is covered with small, ovate, highly carinated scales without any shields. *Mouth* very large; jaws of nearly an equal length; anterior teeth in lower jaw are long, slender, and very nearly upright, the others shorter, few, reflex; a maxillary and palatal series of teeth in upper jaw; fangs longer and stouter than those of any of the Cobras. *Eyes* large, placed high up, and not prominent. *Trunk* round, thick, and beautifully marked, covered with oblong, oval, carinated scales excepting those adjacent to the scuta, which are smooth, broad, ovate, larger, and not carinated. *Color* of head and trunk a yellowish-brown; the back marked with twenty-two or more large, oblong, oval spots, brown in the middle, borders black, and edged with white; some of these spots are separated, but most of them are joined by a narrow neck, or run waving into each other; small black dots, single or two or three together, are sometimes interspersed; a second row of spots adorns the sides similar in color to the first, but smaller and more orbicular in form, each of those on the trunk having a short stem tending obliquely toward the abdomen, and made up of smooth black scales; in the interstices angular black spots are disposed along the verge of the scuta; all these spots become more and more obscure as they approach the tail. *The scuta* are white, glossy, with a membranous striated margin, and many of them are marked with one or two dusky semicircular spots hardly visible near the tail. The *subcaudal squamae* are a dusky yellow, but not spotted.

OBSERVATIONS.

The *color* varies in different individuals, fading perceptibly while in captivity. The number of scuta varies inconsiderably in the different varieties.

This is a rarer species than the Cobra, and much less known.

Second Species—VIPERA ECHIS, has the following varieties:
No. 1. Carinata.
2. Striata.
3. Variegata.

Vipera Echis Carinata.—The only species known in India. Pseudoboa Carinata, *Schneider*, Echis Carinata, *Merr.*, tent. Called *Afaë* in Delhi, also Horatta Pam, Russell. In Sind it is called Kuppur, where it is said to be the most deadly variety found.

Other varieties are found in South America; they seldom exceed 24″ in length, but are generally from 17″ to 20″. It is viviparous, very fierce and aggressive, always ready to attack; springs three or four yards to bite; has long fangs.

Head covered with keeled scales; a pair of very small frontals behind the rostral; nostrils small, round, in a large nasal, subdivided behind the nostrils; sides of the head covered with keeled scales, two series of which are between the eye and the low upper labials. *Scales* much imbricated, in from 25 to 29 rows, strongly keeled, and those in the lateral series have their tips directed obliquely downwards. *Scuta* brown or brownish-gray, with a series of subquadrangular or ovate spots, whitish, edged with blackish-brown; a subsemicircular whitish band on each side of each of the dorsal spots, inclosing a round dark-brown lateral spot; a pair of oblong brown black-edged spots on the crown of the head converge forward; a brown spot below, and an oblique broad streak behind the eye. *Belly* whitish, and more or less numerous round brown specks. Abdominal scuta 149 to 154, subcaudal squamæ 21 to 26.

No. 2. *Vipera Echis Striata* (has two sub-varieties) is a South American variety, and only differs from the preceding in the colors of the skin, it being marked by longitudinal streaks of darker and lighter color from neck to tail. It has no fangs.

Body not thick; length from 25 to 75 centimetres. *Dorsum* covered with 26 to 30 rows of semi-oval lanceolate scales.

It coils itself up in the shape of a figure 8 when preparing to bite. Its movements are exceedingly quick. For effects of the poison, see page 89.

No. 3. *Vipera Echis Variegata*, another South American variety, only differs from the preceding in the variegations of its skin, which are marked by black lozenge-shaped maculæ upon a lead-colored ground, extending down the centre of dorsum from neck to tail.

Fourth Species—VIPERA PSEUDECHIS MAJOR, is very abundant in the highlands.

Length from 75 to 200 centimetres. *Body* 3 to 5 centimetres thick. Throat and *tail* long, latter slender; dorsum greenish-brown, marked with maculæ not of a uniform size, and of irregular shape, covered with 26 to 40 rows of semi-oval lanceolate scales. *Belly* light lead color.

The species which are found at a height of from 500 to 1500 metres above the level of the sea are the most venomous; those occurring higher up being provided with a much less active poison. Its Spanish names are Taya, Taya-Equis, Taya-Rabona.

Seventh Species, VIPERA ACTRACTASPIS IRREGULARIS, has the fangs so long that they almost touch the angle of the mouth.

Eleventh Species, VIPERA LACHESIS BUFOCEPHALUS, called Frog-headed Mapana, or Cabeza de Sapo.

Length from 3 to 6 feet; *head* wider than the thickest part of the body, triangular shaped; the sides of the mouth are covered with 8 large irregular-shaped scales on each side of the upper jaw; the upper anterior edge of the head, from one eye to the other, has its periphery marked by a series of scales folded over each other, forming a sort of ridge, which lies in the same plane as the top of the head, which is perfectly level, and covered with exceedingly minute, imbricated scales, that increase in size towards the neck, and are of a greenish dirty-brown color, like the back of a frog; hence its name. *Belly* of a light lead-color; its back is marked with a row of rhomboidal maculæ of a dark bistre color upon a lighter background, and covered with small round spots.

It has long, stout fangs; it is viviparous. One, which was kept in a small box for a long time in South America, gave birth to 20 snakelets. It is exceedingly abundant in the valley of the River Magdalena.

VIPERA LACHESIS.

Second Variety, VIPERA LACHESIS TRIGONOCEPHALUS.

This variety is not known to have been found anywhere except in Dutch Guiana, where it is called Curucucu, and in British Guiana, Conanaconchi or Bushmaster. Its length varies from 3 to 14 feet; it is supposed to be the largest poisonous snake known; this is the original Trigonocephalus of Dr. Hering; the poison of which is used so frequently in homœopathic practice as a medicinal agent.

The following description is that of a specimen in the possession of Dr. Boericke of Philadelphia.

Length 7′ 7″; abdominal scuta 228 entire, subcaudal squamæ 37 pairs (making 265); *body* covered with 26 rows of scales, the central row over the axis of the dorsum most highly carinated, each successive row being less so down towards the sides; scales diminish slightly in size toward the neck, gradually and much more so towards the tail, on the point of which they are very small; tip of tail 1½″ in length, and terminates in a sharp, horny point, ¼″ in length; *head* considerably depressed at the eyes, but rises towards the rostrum, the extreme point of which is as high as the occiput; *head* covered with very small scales, gradually increasing in size towards the neck; *rostrum* covered by one large triangular shield; upper periphery of the mouth is sheathed by 8 shields, precisely the same as those described in the *Vipera lachesis bufocephalus;* the largest shield is between the eye and the nostril; *nostrils* gaping, slightly turned backwards, infundibuliform; *body* marked with a zigzag line of semi-rhomboidal maculæ, having the angles on one side opposite the points on the other; the rhomboidal maculæ are broken in several places along the line of the dorsum; the maculæ and upper surface of the tail are of a uniform color of very dark bistre, approaching to black; *scales* on the back are shaded with the same dark color on the front side, and of a dirty, yellowish-light brown color on the back side; *belly* of a murky-yellowish white in the centre, tinged with a lead-colored shade on the sides; under side of the jaws, throat, and neck, covered with medium-sized scales, regularly disposed in diagonal rows.

The lowest angles of the rhomboidal maculæ (which approach the nearest to the sides of the scuta), terminate abruptly, without completing the angle, thus leaving a break in the mark or rhomboid; this is inclosed by a small semi-rhomboid, the angle of which terminates just above the upper edges of the scuta; the periphery of these smaller maculæ is somewhat indistinct in places. It has long, stout fangs, and the disposition of the buccal parts is identical with that of the *Crotalus fasciatus.* See page 81.

No. 3. *Vipera Lachesis Niger*, called Mapana fina, and when an individual of this variety has the tip of the tail terminating in a horny point, it is called Mapana de Puya (pua).

Length from 3′ 6′′ to 4′ 6′′; diameter always proportionate to its length; *head* hastate, triangular, black on its upper surface, greenish-white on its periphery and under the throat; *dorsum* blackish, marked in angular, irregular-shaped lines, and maculæ on a lighter background, which shades down gradually into a greenish-white on the belly; *trunk* tapers gradually towards the tail, and also towards the neck, which is very long and rather slender.

It has fangs that are long and slim. Its movements are exceedingly graceful and rapid; possibly, there is not another snake known in the world which is more aggressive or quicker to bite, or in its movements.

The female when it has just given birth to its brood, if attacked or threatened, receives into its mouth the snakelets, which come out from their hiding-place again as soon as the danger has passed. This same fact is mentioned by several naturalists as having been observed in other varieties.

No. 4, *Vipera Lachesis Acuaticus*, and No. 5, *Vipera Lachesis Variegata*, abound in marshes and lagoons; the former coils itself in the branches of the trees at the edge of a river or brook, during the hours of extreme heat, and lets itself drop down into or upon a passing canoe or boat, much to the consternation of the "bogas." Their poison is not so deadly as that of the *Bufocephalus*.

No. 6. *Vipera Lachesis Dryiophis*, is marked like both No. 4 and No. 5, but it is always coiled upon the trees at the edge of the water and only lets itself drop when disturbed or molested.

No. 7. *Vipera Lachesis os Flavus*, Boguidorada or Gilded mouth. This is a common species in the valley of the Magdalena and Valle Dupar region; it also occurs in Venezuela. In size, and the maculæ on its dorsum, it bears close resemblance (in fact in many individuals is identical), with the Bufocephalus, but the throat and body are rather more slender and of a more graceful shape. The sides above the edges of the scuta are of a lighter color, and the edges of the mouth and throat are of a golden-yellow hue, hence its name. Its poison is very deadly and its bite much feared, as the Curers consider a cure of it a great test of their powers and proof of the efficacy of their secrets. They lose, however, many cases, but save their reputation by attributing the unfortunate termination of the case to a tithe (every tenth case) due to "La Suerte," their god of Chance, whose assistance they invoke upon undertaking to perform every case of cure.

No. 12. *Vipera Calamaris Venenosus*, occurs at an altitude of 300 metres and upwards above the level of the sea. It is never found in the lower parts of the great rivers.

Length from 15 to 40 centimetres; *body* of nearly a uniform thickness; *tail* short and more highly prehensile than that of any other variety known; *dorsum* covered with 18 to 20 rows of semiovate, lanceolate scales; *mouth* and *head* disproportionately large and broad; it is provided with two very large, stout fangs; disposition of buccal parts same as those of *Crotalus fasciatus*. *Color* (in the male) dark brown on the back; top of head of a darker shade; *belly* of a creamy white tinted with a roseate hue on the sides, and of a darker hue on the subcaudal part. The *female* is a little larger, longer and thicker than the male, and has its dorsum marked with nine longitudinal stripes (one dark and the adjacent ones of a lighter shade), running from the neck to the anal region.

It winds its tail around any small object close to the ground, and will spring to a distance of two and a half or three yards with ease. A peculiarity of its bite is that the wounds made by the fangs *are almost always diagonal* to the axis of the bitten limb. This fact makes its bite easily recognized by the Curers. The popular belief in Northern South America is that its poison kills before twenty-four hours have expired after its injection, hence one of its Spanish names, Veinticuatro (twenty-four), or Patoquilla.

A second variety is *Vipera Calam. Venen. Rubrum*, called Candelilla or Flame snake, from its color. It is found in the States of Antioquia, Boyaca, Bolivar, and Magdalena (United States of Colombia).

Length from 25 to 40 centimetres; diameter 1½ to 2 centimetres; *body* of nearly a uniform thickness, covered with very minute lanceolate scales; *color* of head and body a fire-yellow, of a darker shade on the head and dorsum, and of a reddish-white on the belly; *buccal parts* identical with those of the *E. Cuprocephalus*.

Nos. 14, 15, 16, and 17, are Australian species, of which no description is given.

No. 18 is the Puff Adder of Africa, which when provoked swells or puffs itself out like a frog. The Hottentots are said to kill this snake by spitting tobacco-juice in its mouth.

A second variety of No. 18 is *Vipera Clotho. cornuta*, Das Adder or the River Jack, sometimes taken for the Cerastes.

It is also called the Plumed Viper or Hornsman.

No. 19. *Vipera Cerastes Caudalis*, the Cerastes or Horned Viper, is one of the most vicious-looking reptiles in existence.

Its general *color* is that of a frog, variegated with irregular-shaped bistre-colored maculæ and spots on its dorsum; *length* from two to three feet; *trunk* very thick and slightly carinated; *head* cordate, and the *rostrum* is provided with two short, upright, tentacula-like appendages called horns.

No. 20. *Vipera Cench. Piscivorus*, is also called Cottonmouth or Water Moccasin. This snake's poison sometimes produces death and at other times it does not, consequently it is little feared.

No. 23. *Vipera Acanthophis Antarctica*, is also called Death Adder or Death Viper, in Australia.

Other snakes occur which are known to belong to the genus Viperidæ, but those given are the only varieties yet classified, making a total of 25 species, composed of 69 varieties, &c.

The *Amphisbœnœ* of Buffon's classification has been excluded from the family of Ophidians by later naturalists, as belonging more properly to that of the Earth-worms, but some of the species of this genus are found in South America, particularly in the forests of the great Atrato Valley and of the Isthmus of Darien, that are known to be exceedingly venomous.

One variety, occurring also in Brazil, is the *Amphisbœnœ Vermicularis* (Wagler, Mure), signifying "worm snake, that moves with equal facility forward or backward," is called Tatacoa, Culebra ciega, Tatacua, Sin ojos, Culebra de dos Cabezas, Culebra gusano. The variety described by Dr. Mure is, viz.:

Length 50 to 80 centimetres; *tail* obtuse; *body* of nearly a uniform thickness, variegated by small square maculæ, divided from each other by entire, narrow rings, which embrace the whole body; *head* small, slightly compressed; *eyes* hardly discernible; *teeth* stout in proportion to its size, slightly curved backwards; *belly* whitish with a roseate shade at the sides; *sides* and *dorsum* marked indiscriminately with small dark spots; *color* light lead, maculæ and rings a dark brown or nearly black.

In the district of Guamocó (United States of Colombia), in the State of Antioquia, a snake abounds which varies in length from four to ten feet, called Birri. It has large fangs, is slow in its movements, like the *Crotalus horridus*, and its skin is variegated somewhat similar to the *Lachesis bufocephalus*.

Never having been able to visit that district, I solicited a celebrated Curer to bring me a specimen alive, but never was able to procure one.

Its poison is exceedingly deadly, and so peculiar in its action, that shortly after injection into the blood, large masses of flesh rot and drop off, and this continues till the bone is laid completely bare and a large vein is reached, when death ensues immediately. This fact originates the name of Podredora, or Rotter, it has in some places.

I only knew of one case of its bite in which death did not ensue, but the flesh had sloughed away up to the knee, leaving the upper surface of the foot and lower leg nearly bare, and one entire ulcer, which healed in about thirty days by taking a tablespoonful of *Bilis. Acrochordon Chocoe* (ten drops in ten ounces aq. dest.) daily.

All cases of its bites in Guamoco are now (1873) cured by a preparation of its gall. As yet no case has been known to terminate fatally since this remedy has been used in this district, where its use was introduced under my directions by a miner in the year 1865.

Where references have been made to some of the physical peculiarities of different varieties, the following remarks are intended to apply in a general sense. Snakes in captivity are less venomous than in a state of nature. The *Crotalus* in tropical forests emits so strong a smell of musk as to impregnate the surrounding atmosphere with it, as though to warn one of its proximity; this fact has misled many observers.

When a snake in the forest loses a fang, or both of them, twenty-four hours is sufficiently long for the new fangs to come forward and fix themselves firmly to the superior maxillary bone; as I have observed in four or five cases in South America, where I have tempted them to seize small pigeons and lizards; would allow them to get a firm hold, and then tear the prey from their grasp with violence, wrenching off one or both fangs; after allowing twenty-four hours to elapse thereafter, I have caught the snake in his hole, and invariably found the new fangs firmly anchylosed to their maxillaries.

The most intelligent Curers all entertain the belief that in the tropics no more than twenty-four, or at most forty-eight hours in some cases, are necessary for a fallen or broken fang to be replaced by a new one.

The following varieties are not placed in any of the preceding tables, as their classification is not known.

Found in South Africa.

Coristodon concolor.
Cyrtophis scutatus.
Dasypeltis, { Inornatus, Scaber.
Elapomorphus capensis.
Heterolepis capensis.
Homalosma arctiventris.
Lamprophis, { Rufulus, Aurora.
Monopeltis capensis.
Onychocephalus Delalandii, 6 sub-varieties.

VIPERIDÆ.

Sepedon rhombeatus.
Temnorhynchus Sundewallii.
Telescopus semiannulatus.
Thelotornis capensis.

Trimerorhinus rhombeatus.

Varieties classified, 37
" not " 21=58

Found in Australia.

Acanthophis antarctica.

Cachophis, (4) { Krefftii. Fordii. Harriettæ. Blackmanii.

Denisonia ornata.

Hoplocephalus, (18) { Curtus. Superbus. Ater. Variegatus. Stephensii. Pallidiceps. Gouldii. Spectabilis. Coronoides. Mastersii.

Hoplocephalus, (18) { Signatus. Temporalis. Ramsayii. Minor. Nigriceps. Nigrescens. Nigrostriatus. Maculatus.

Petrodymon cucultatum.

Vermicella, { Annulata. Lunulata.

Varieties classified, 47
" not " 27=74

TABLE OF THE DIFFERENT SNAKES FOUND IN THE UNITED STATES OF COLOMBIA,

*Noted under the Names by which they are commonly known.**

STATE OF ANTIOQUIA.

Venomous.
- Taya-équis (nobis). ⎱ Of these two varieties, the second is possibly the most deadly.
- Birri, " ⎰
- Torcoral, or Corcorá. A semi-ophidian, half snake and half lizard, having a crest upon its dorsum, and four short legs like the Guatáca.
- Equis (four or five varieties), (nobis).
- Cascabel (Crot. fasciatus), (nobis).
- Candelilla (nobis).
- Vibora " sundry varieties.
- Coral "
- Lomo de machete (nobis).
- Ciega, Tatacóa (amphisb. verm.), (nobis).
- Rabo-ají (mapana), (nobis).
- Arará.
- Voladora.
- Yerga.
- Taya (nobis).
- Yaruma (this is probably the "Birri." See page 103).

---

* Many of these names are taken from the geography of the United States of Colombia. Bogota, 1862, 2 vols.

Non-Venomous.
- Reina.
- Coclí.
- Pitora.
- Paloma.
- Guarda-camino.
- Viní.
- Boa C. (sundry varieties).
- Boba (Python), (nobis).
- Verde (nobis).

Besides these there are many varieties in the woods and virgin forests of the State, which are unknown.

STATE OF BOLIVAR (*nobis*).

Venomous.
- Birri (only found in the District of Guamocó).
- Mapana (all the varieties).
- Cascabel.
- Patoquilla, or Veinticuatro.
- Lomo de machete.
- Torcoral.
- Candelilla.
- Vibora (sundry varieties).
- Toche (lives on dry land).

Non-Venomous.
- Toche (aquatic variety).
- Cazadora (ambiguous; some venomous, others innocuous), (some individuals occur three metres long).
- Guata.
- Bejuco.
- Boa.
- Verde.

STATE OF BOYACA.

Venomous.
- Taya (nobis).
- Cascabel.
- Coral (one variety of this is said to occur called "the two-headed").
- Taya-équis (nobis).
- Equis "
- Tatacoa "
- Sapa "
- Tigre "

Non-Venomous.
- Mamadora (nobis). A snake often found sucking milch cows; and cases have been known where a woman has been asleep, with a child nursing at one breast and a snake sucking the other!
- Verde.
- Galana.
- Bujio, or Buio.
- Paja.
- Bejuco.
- Boa (four varieties).
- Traga-venados.

STATE OF CAUCA.

Venomous.
- La Verrugosa.
- Cascabel (Crot. horr.), (Crot. fasciatus).
- Equis.
- Coral.
- Veinticuatro, or Patoquilla.
- Tigre, or Mapana Tigre (Lachesis; nobis).
- Rabo de chucha. "Dry tail," or "Yellow tail."
- Cachetona.
- Yaruma (Birrí; nobis).
- Voladora.

IN COLOMBIA.

Non-Venomous.
- Sobre-cama.
- Arco.
- Coclí.
- Bigola.
- Papagayo.
- Petacona, or Boa.

Territory of Caqueta.

Venomous.
- Equis.
- Coral.
- Mapanare (Mapana-Lachesis; nobis).
- Sapa   " Sapo;   "
- Tigre   " Tigre;   "
- Voladora.

Non-Venomous.
- Boa.
- Traga-venado.
- La Matiguaja.
- Loro Estrella.
- Las-pacas.

State of Cundinamarca.

Venomous.
- Equis (sundry varieties, Taya-équis).
- Coral.
- Mola.
- Teti, or Tití.
- Víbora (sundry varieties).
- Tatacóa, or "two heads" (Amphisb. vermicularis).
- Mapanare.
- Labrancera.
- Candelilla (nobis).
- Toche   "

Non-Venomous.
- Mamadora (nobis), (the "Sucking Snake.")
- Rayona.
- Sabanera.
- Verde.
- Petaca.
- Traga-venado (Boa C.). These last two are found in the great plains of San Martin.
- Cazadora.
- Tiro.
- Negro.
- Dormilona, or Boba.

State of Magdalena.

Venomous.
- Cascabel.
- Taya.
- Coral.
- Guata.
- Mapana (sundry varieties; nobis).

Non-Venomous.
- Boba.
- Boa C.

State of Panama.

Venomous.
- Verrugosa (Acrochordon Chocoe; nobis).
- Equis.
- Víbora (sundry varieties).
- Coral (Elaps Corallinus; Mure).
- Sin ojos (Amphisb. vermicularis; Wagl.).

Non-Venomous.
- Cazadora (Coluber; nobis).
- Bejuco (six varieties).
- Boba.
- Boa, or Petaca (Boa C.; Cuvier).

STATE OF SANTANDER.

Venomous.
- Cascabel.
- Taya.
- Coral.
- Mapanare (Mapana, Lachesis).

Non-Venomous.
- Huertera.
- Voladora.
- Cazadora.
- Guata.

STATE OF TOLIMA.

Venomous.
- Taya.
- Taya-équis.
- Equis.
- Cascabel.
- Coral.
- Taya-rabona.
- Vibora (sundry varieties).
- Mapanare (Mapana).

Non-Venomous (Doubtful).
- Dormilona (Boba; nobis).
- Sabanera.
- Tiro.
- Verde.
- Negra.
- Toche de tierra fria (highland or lowland varieties).
- Rayona.

Many varieties, and even species, undoubtedly exist in the virgin forests of Colombia, which are as yet not known to the

Curers; and it is an undoubted fact, that on the Great Plains of Casanare and Saint Martin, so many species and varieties do exist that it is probable that the preceding tabular list embraces only about *one-third* of the existing number. Such individuals as have been seen on the Great Plains seem to possess every possible variegation of marks, maculæ, and colors in the skin; and it is easy to trace every step in a progressive development through individuals which have no traces of limbs up to the Torcoral, which bears strongly marked characteristics of the Ophidian on the one side, and of the Guataca (the lowest species in the Saurian family) on the other, thus affording an argument in support of the Darwinian theory of progression of species.

Toxical effects of the *Elaps Corrall.* poison are less violent than those of the *Crotalus* or *E. Cuprocephalus*, but are none the less worthy of study, as its action is very marked on particular organs and viscera. The respiratory organs, those of deglutition and the brain, are first affected. Sudden and excruciating pains present themselves in a particular organ or viscera, or muscles, are felt for a short time, and as suddenly cease, to reappear in some other organ or part. A flow of venous blood,* in coagula, presents itself from the mouth, eyes, ears, urethra, and from the wound or bite; pains in the abdomen ensue; there is extreme sensibility to the touch in the whole body; an almost total extinction of the pulse; colic, diarrhœa; tremors in the muscles; excessive thirst and subsequent œdema in the bitten part, succeeded by vomiting and fainting fits. Death does not always ensue, and this poison is undoubtedly less virulent than that of Crotalus.

---

* The mixture of poison with the blood causes the latter to change its color to a very dark purple-red, *in all cases* of poisoning by that of the Crotalus type.

The following formula expresses at a glance its entire action:*

$$\frac{s\Omega}{1} \cdot \frac{I^{\pi+\lambda} \quad J^{+\frac{\omega}{2}} \quad J^{\bar{\omega}}_{\frac{\omega}{2}} \quad O^{\lambda}_{1-3} \quad Y^{\lambda}_{1-6}}{R^{\delta}_{1-3} \quad P^{\lambda}_{1-3} \quad G^{\varphi}_{1-5} \quad E^{\lambda}_{1-6} \quad A^{\varphi}_{1-6} \quad \left(ZXL_{\infty}\right)^{\delta+\lambda} \quad Q^{\pi}_{1-6} \quad M^{\delta}_{3-6}}$$

It reads thus,—*Numerator:* The intellectual faculties are clouded by illusions, accompanied by eruptions on the skin and a feeling of insensibility to the touch in certain parts. The preceding symptoms developed throughout the entire proving. Functions of the nervous system are excited, sensation in some increased, and afterwards diminished.

Sense of hearing much diminished from first to third day; clearness of vision diminished from first to sixth day.

Denominator: Respiratory apparatus very strongly affected from first to third day; feeling of insensibility in penis, first to third day; difficult deglutition, first to fifth day; digestive functions deranged, first to sixth day; abdominal regions strongly affected, first to sixth day; organs of the throat and abdomen, and the entire trunk, affected during the whole proving, particularly insensibility of the organs of the trunk; eruptions in the cutaneous tissue from first to sixth day; matrix notably affected from third to sixth day.

The fraction expressed in signs at the left hand means: Development or manifestation of the principal symptoms on the right side of the body during the entire proving.

Toxical effects of the poison of the *Vipera lachesis bufocephalus* are: Convulsions; great pain in the bitten limb; intensely feverish pulse; intense thirst; flow of a dark blood,

---

* From Mure's Pathogenesie Brésillienne.

semi-coagulated, from the nose, ears, and rectum; veins of the conjunctiva are surcharged, so that the eyes have the color of raw beef; tremulousness and subsultus tendinum; discoloration of the bitten part; the blood is coagulated near the wound, and will only flow from it after deep incisions have been made just at or above the wound made by the fangs, and the limb has been bathed repeatedly in hot water, rubbing it down towards the bitten part so as to force out the coagulated blood. This poison causes death in from three hours to three days, according to the "condition of virulence" of the poison.

The Boqui Dorada, Yellow-mouthed Viper, or *Vipera lachesis os flavus*, is possessed of a very virulent poison, which causes as many deaths as that of any other variety in the valleys of the River Magdalena and its tributaries. Toxical effects are: Flow of blood from the nose and mouth; evacuations of fæcal matter, with clots of dark-colored blood, or scanty, suppressed evacuations; heaviness of the head; stupor; sensation of oppression in the chest and lungs; repeated blows upon the tympanum (like those made by a hammer); vertigo; loss of sight; intense pains in the spinal column and the shoulder-blades; cramps; colicky pains in the abdomen; throbbing pains in the bitten part, increasing with the œdema; discharge of bloody urine; pulse has a slow, heavy beat; rheumatic pains in the muscles in twinges; blue-black spots under the nails of the fingers and toes; lips discolored; eyes bloodshot; chills, followed by tremors in the whole body, while the skin is flushed and indicates a highly feverish condition; flow of a greenish, sanguinolent foam from the mouth. This latter symptom invariably precedes death, and continues for some hours after it has ensued. Death ensues in from three hours to three days.

The poison of *Vipera lachesis niger* always produces a shock, which throws down the person bitten with great violence; an immediate and violent flow of blood ensues from eyes, mouth, nostrils, urinary canal, and from under the nails of the fingers and toes; veins of the conjunctiva are intensely injected; suspension of the urine ensues, followed by violent fugitive pains in the bitten limb, and intense cephalalgia. Death ensues in from one to twelve hours. A Curer informed me that he had known persons bitten by this snake to die in five minutes; but the only case to which I can attest is that of a dog, bitten by one of these snakes, that died in less than twenty minutes.

The poison of the *Vipera Calamaris venenosus* develops the following toxical symptoms, viz.: Flow of blood from mouth, nose, and ears; suspension of urine; subsultus tendinum; pains in the renal region; and lastly, ungovernable madness. This last symptom is developed in almost every case where the poison is active.

The toxical effects of the poison of the *Vipera pseudechis major* are: Atrocious pains in the entire body; flow of blood from mouth, nose, and ears; scanty or suppressed urine; constipation; rheumatic pains, which recur for months and even years, in cases where the bitten persons recover.

The poison of the *Vipera echis striata*, when it does not cause death, is said to develop pains of a rheumatic nature, which increase and diminish according to the age of the moon. If this symptom is constant, it is common with a similar one developed by the *Crotalus horridus* poison referred to on pages 120–121.

The toxical effects of the poison of the *Vipera Calamaris venenosus rubrum* are as follows, viz.: One or two hours after its injection, the surface of the bitten limb is covered with a great number of small red vesicles; when these attain the size of a grain of pearl barley, they burst, and discharge a corrosive, yellowish, ichorous liquid, accompanied by a burning, itchy sensation, and followed by œdema of the limb; acute pains extend up the limb; excessive thirst; inflammation of the conjunctiva; pulse 140 to 160; dislacerating pains in the head; the pustules increase hourly in size till they attain an inch or more in diameter; and if the bitten person does not die, these are so many wellnigh incurable ulcers.

I have cured two cases of these ulcers: one with the gall of the centipede, and the other with gall of the Acrochordon chocoe, whose poison produces a similar eruption and ulcers.

The Curers say that this snake possesses phosphoric properties in its skin, which make it visible in the dark; hence its name Candelilla, which means literally, "an insignificant little light."

Toxical Effects of the Poison of the Crotalus Cascabella of Brazil.

The wound was inflicted in one of the fingers of a man's hand. The snake measured two and a half metres in length.

Symptom No. 1. Swelling of the hand, and drops of blood escape from the wound.

Pain in the palm of the hand, which extends up to the wrist.

At the expiration of one hour from the injection of the poison, the hand is exceedingly swollen, accompanied by a sensation of cold, which is also felt in the lower extremities.

Pulse strong; increases at intervals from 110 to 140 per minute.

EFFECTS OF POISON OF CROTALUS CASCABELLA.

No. 5. Sensation of plenitude in the jugular veins, which is soon felt in the sides of the throat and nape of the neck.
Feels a blur before his eyes.
A crawling sensation in the face.
Expiration of 1 hour and 30 minutes. Pain and œdema extend from the hand nearly up to the elbow.
The whole body seems to fill itself up.

No. 10. Expiration of 1 h. 20 m. Visible tremors in whole body.
Unpleasant sensation in the head.
Pulse accelerated.
Feels it difficult to move the lips.
Inclination to sleep.

No. 15. Sensation of constriction in the throat.
Swelling of the hand continues.
The entire arm is swollen and very painful.
Expiration of 1 h. 38 m. He feels cold, and desires to cover himself.
Expiration of 1 h. 48 m. Feels a pain in the œsophagus, which extends to the stomach and abdomen.

No. 20. Sensation of cold in the feet.
Expiration of 2 h. 5 m. Feels it difficult to speak.
Expiration of 2 h. 25 m. Swallows with difficulty.
Anxiety.
Copious perspiration in the chest.

No. 25. Expiration of 2 h. 50 m. Swelling of the arms; slight epistaxis.
Inquietude and anxiety increases.
Pulse 96.
Expiration of 3 h. 4 m. General perspiration.
Involuntary groans and moaning.

No. 30. Feels exceedingly downcast.
Pulse 100.
Expiration of 3 h. 15 m. Pains in the arms; inquietude.
Expiration of 3 h. 30 m. Pulse 98.
Face much flushed.

No. 35. Flow of blood from the nose.
The whole body is flushed and red.
The blood starts from a pustule in the armpit.
Expiration of 4 h. Body a much darker red.
Unbearable pains in the thorax; patient excessively prostrated.

No. 40. Constriction in the throat; respiration difficult.
Expiration of 4 h. 30 m. Pulse 104; great pain in the surface of the whole body.
Salivation.
Expiration of 5 h. 30 m. Pulse 104.
Has a feeling of stupidity.

No. 45. Abundant discharges of urine.
Saliva thick, dark-colored, viscous; spits with difficulty.
Muscular prostration.
The pains cause frequent groans.
Respiration easy.

No. 50. The skin is moist.
Expiration of 7 h. Somnolency and groans.
Pains in the breast and arms cease.
Sensation as if a knot in the throat.
Copious urinary discharges.

No. 55. Deglutition difficult.
Saliva whitish, viscous.
Flow of sanguinolent serosity from the nostrils.
Expiration of 9 h. 15 m. Profound sleep.

The preceding table indicates that the person bitten was of a strong physique, and the venom did not have its deadly principle fully developed. In extreme cases, death ensues in from one to four hours—sometimes sooner.

Immediately after the injection of the poison, there is a sudden increase in the rapidity of the circulation of the blood; sharp pains in the chest, throat, and limbs; much œdema of the part bitten; sensations of flushes of heat in whole body, followed by ice-cold chills; abundant epistaxis; discoloration of the finger and toe-nails, and flow of blood from under them; blood flows from the gums, roof of the mouth, and from the urinary canal; the pulse rises suddenly to 140 or 160, and as suddenly lowers to 75 or 80, to rise again to 100 or 105; surface of the skin becomes red; soon after it changes to a purple hue; oppression in the chest is felt; great muscular debility ensues, which is superseded by a state of coma; patient rouses slightly, complains of inability to distinguish objects around him; relapses into the comatose condition; a bloody or dark-colored or greenish froth is noticed on the lips, and death ensues immediately thereafter.

Dr. Weir Mitchell's[*] conclusions as to the great similarity between the action of one serpent's venom and that of another of a different genus, are quite contrary to the experience of every person who has had to cure many cases of snake-bite; so much so, in fact, that, as a general rule, an experienced Curer will detect indications in the person bitten which will enable him *to name the kind of snake* which caused the wound.

Experiments, properly initiated and conducted, will place this question beyond the domain of discussion, however, and will enable us to know positively the field of action of each

---

[*] Researches, &c., page 98.

one of the poisons, as far as possible to determine it with the means at our disposal.

In another place I have referred to a fact which has, possibly, escaped the attention of observers hitherto; this is, that the venom which a reptile secretes in a state of nature, and that which it secretes in a state of captivity, possess different degrees of virulence and develop somewhat different symptoms. My own observations and studies lead me to attribute this difference to the existence of an electro-magnetic element in all snake poisons, which is fully developed in a state of nature at the period in which the reptile is most constantly in motion, and which is at its minimum when it is in a state of torpidity, or of little exercise, as in captivity.

This will also explain why Dr. Weir Mitchell's and others' observations induce them to adopt the belief that the *Crotalus* does not emit a strong smell of musk; while, in a state of nature, this odor is so strong as to be exceedingly offensive.

SNAKE POISONS,

THEIR NATURE, ANALYSES OF THEM, AND SOME INTERESTING FACTS DERIVED FROM THE STUDY OF THEIR ACTION.

Figure 1 represents the microscopical appearance of human blood: *a a a* is a front view of the globules, and *b b b* shows an edgewise view of them. The white background between the globules represents the plasma or liquid or vehicle through which the blood-globules are carried in the circulation.

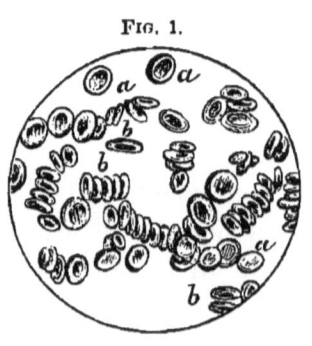

Fig. 1.

The plasma is composed of water, containing in solution a great number of different substances. The presence of albumen, fibrin, various fatty substances, some of which contain sulphur and phosphorus, a great number of salts, such as the chlorides of potassium and sodium, chloro-hydrate of ammonia, the sulphates of soda and potassa, the phosphates of soda, lime, and magnesia, the carbonates of soda, lime, and magnesia, and of alkaline salts, formed by fatty acids and lactic acid, has been detected in blood. This plasma contains also several gases in solution: oxygen, carbonic acid, and nitrogen, which are derived from the action of the air in the lungs. It has a peculiar mawkish taste, characteristic in some animals, and always exerts a well-marked alkaline reaction, which appears to be an essential part of its nature, for animal life ceases when, by direct injections, the blood can be made acid.

In a healthy man 100 parts of blood contain on an average 79 parts of water, 1 part of mineral salts, and 19 of albuminous substances, known by the name of *hæmatosin*, which proportions vary greatly with the state of health. In the blood of birds the relative quantity of water is generally somewhat smaller than in a man, while it is greater in that of the batrachian reptiles and fishes. As much as 98 per cent. of water has been found in the blood of a frog.

Besides the red globules, other colorless globules are seen, the number of which vary under different physiological conditions; these are more apparent in the *serum* when by coagulation the blood has separated itself into the former, a yellowish and transparent mass, and into a gelatinous mass of a deep red color, called *clot, coagulum* or *crassamentum*. The phenomenon of coagulation is produced by the fibrin, which remains in solution so long as the blood is under the influence of the vital principle; but separates from it when it is removed from the animal economy, carrying with it the blood-globules, in the same way that soluble albumen, used for the clarification of a muddy liquor, carries down the corpuscles which exist in it as soon as it is coagulated by heat. If, instead of allowing the blood to rest, it is beaten with rods, the fibrin still coagulates, and forms whitish and elastic filaments, which adhere to the rods, the blood-globules not being included, because they are detached by the agitation of the fluid. Defibrinated blood no longer coagulates.

It is easy to demonstrate with the blood of frogs, the globules of which are too large to pass through filtering-paper, that fibrin is really in solution in the serous liquid, and does not constitute any part of the globules, as was long supposed.

It is sufficient to pour upon a filter, previously moistened, the blood of a frog, at the moment of its extraction, to show that a portion of the liquid passes through the filter before the commencement of coagulation; and after collecting this portion in a watch-glass the microscope will exhibit in it, after a short time, a colorless clot, which may be made visible by collecting it on a needle. This experiment does not succeed in human blood, nor in that of other mammiferæ, because the fluid is more viscid, and the globules are sufficiently small to pass through the paper.

Serum is a yellow, slightly viscid fluid, of a density of 1.027 to 1.029, with a slightly saline taste, and coagulates at about 168.8°, which is a property of the albumen element.

Several saline substances prevent coagulation, for example, sulphates of soda, the chlorides of sodium and potassium, nitrate of potassa, borax, &c., and the proportion of these salts must be about one-sixth of the weight of the blood.

Dilute mineral acids also prevent the coagulation of blood, but impart to it an oily consistency. A temperature of 86° to 104° Fahr. appears to be the most favorable for coagulation, while cold retards it considerably.

Healthy human venous blood yields:

Coagulum,	13.0
Serum,	87.0 = 100.0

Coagulum, {	Fibrin,		0.30 } 13.00
	Globules, {	Hæmatosin,	0.20
		Albuminous matter,	12.50

Serum, {	Water,	79.00
	Albumen,	7.00
	Fatty substances,	0.06
	Various salts with mineral bases,	0.94

100.00

Hæmatosin has not yet been extracted in a state of purity, and has not been obtained crystallized. It forms after separation from the other elements of the blood a blackish-red amorphous mass, tasteless, inodorous, and insoluble, when cold, in alcohol, water, or ether.

It dissolves readily in alcoholic solutions of potassa, soda, and ammonia, which it colors intensely red. It contains, more or less, 10 per cent. of sesquioxide of iron, which appears essential to the existence of the globules.

NATURE AND ANALYSIS OF THE POISONS.

Moquin-Tandon* says that the poison of a Crotalus is of a greenish hue, but I have often extracted it from the South American varieties from a greenish to a yellowish tinge; it is neither acid† nor alkaline; is precipitated in water, in which it preserves its viscidity for some time, but ultimately mixes with it. When dried on a plate of glass it looks like a layer of gum full of cracks. It is insoluble in sulphuric and hydro-

---

* Medical Zoölogy, by Moquin-Tandon.

† Dr. Mitchell's tests prove the Crotalus venom invariably acid.

chloric acids, which only reduce it to a liquid paste. In nitric acid it acquires a yellowish tinge, otherwise it appears to be affected identically with that of the two former. Vegetable acids, alkalies, and oils do not dissolve it; when heated it does not melt, but swells up and becomes thick; placed in contact with flame it does not ignite. The active principle of the European viper (*Pelias Berus*) poison is called *Echidnine* or *Viperine* ;* a colorless, transparent, rather thick substance, resembling varnish, without smell or flavor; this principle is a neutral unstable substance representing the ptyaline of saliva. It does not redden the tincture of turnsole nor turn the syrup of violets green; it contains nitrogen; dissolved in a solution of caustic potash the hydrated binoxide of copper turns it to a beautiful violet color. This also occurs with gelatin and albumen. It seldom kills horses, sometimes sheep and cats. A pigeon dies in from eight to ten minutes, a sparrow in five minutes, and the Abbe Fontana said that it took two grains to kill a man. The merit of Fontana's experiments may be judged from the fact that he made six thousand, in which four thousand animals were killed and three thousand vipers used.

Poison of the Crotalus Horridus.†

According to Mitchell the venom of the Rattlesnake is composed of:

1st. An albuminoid body, called *Crotaline*, not coagulable by heat at 212° Fahr.

2d. An albuminoid compound, coagulable at 212°.

3d. A coloring matter, and an undetermined substance; both soluble in alcohol.

---

* Analysis of Viper Poison, by Prince Lucien Bonaparte, in 1846.

† Researches on the Venom of the Rattlesnake, by S. Weir Mitchell, M.D.

4th. A trace of fatty matter.

5th. Salts, chlorides, and phosphates.

The color of the venom varied from *pale emerald green* to *orange* and *straw* color.

When the poison had remained long in the gland, it was deeper in hue than when its ejection followed rapidly upon its formation. In eight observations on record, the venom reddened litmus paper more or less distinctly. *It was uniformly acid,* and this reaction was common to all species of poison, whether moist or dry, dark-colored, or pale in tint.

The reaction of the mucous membrane of the mouth was almost as constantly alkaline as that of the venom was acid. It was found that litmus, reddened by the venom, became blue again when left in the serpent's jaws; but although the acid was neutralized, the poisonous properties of the fluid remained unaltered.

In the numerous experiments made by Mitchell, where the exhausted venom-gland was mixed with water, and introduced into the wounds of various animals, it was ascertained that all the pigeons, except one, escaped; also the rabbit; all the reed-birds died: these latter birds are remarkably susceptible to Crotalus venom, and will frequently die from a quantity of poison so minute, that it would be hard to conceive of its power to destroy life, had not the experiment been actually made. Thus half a drop will often kill a reed-bird in a minute or two, and one-eighth of a drop will prove fatal after a lapse of from two to eight hours; so that it is probable that even a smaller quantity would be sufficient to destroy life.

The secretion in the Crotalus from the gland is *constantly acid,* and in the viper *neutral;* while the saliva of the parotid, in all animals yet examined, is unchangeably *alkaline.*

Saliva is a secretion of rapid formation, and appropriated to specific mechanical and chemical purposes within the economy; the venom-fluid is slowly elaborated, slowly reproduced when lost, and destined to end in the body which produces it.

Its singular nature as a ferment, poisonous to other animals as well as to the reptile in which it is secreted, constitutes a distinction which forbids the physiologist to regard it as in any true sense a salivary secretion, or its forming organ as a salivary gland. Extensive experiments with chemical reagents, as well as with extremes of heat and cold, have unfolded the singular energy of this poison, and the almost inconceivable tenacity with which its powers are preserved.

The dried venom retained its potency after two years of climatic changes; the fresh poison, although prone to partial decomposition, would remain active after a sojourn of several weeks in an atmosphere of 65° to 70° Fahrenheit.

Mixing the venom with water, and freezing it, and keeping it for half an hour at 4° above zero, did not in the least diminish its power; nor did a temperature of 212° Fahr. destroy its virulence.

In eight experiments Dr. Mitchell has shown that:

1. The coagulum produced by heat is always innocent.

2. The supernatant fluid is uniformly poisonous. The cases died with the usual rapidity. A fact which leads us to infer that the venom loses no power by being heated to a certain degree.

The animals which perished from the injection of the *boiled venom* exhibited very trifling local evidences of the action of the poison. Dr. Mitchell is unable to offer any plausible explanation of this curious deficiency. (Were he a believer in the dynamic origin of disease and the action of remedies, he would be better able to comprehend this phenomenon.)

The mingling of the venom with alcohol, sulphuric acid, muriatic acid, ammonia, chlorine water, iodine, soda, and potassa, in a diluted or undiluted state, did not destroy the virulence of the venom.

When it was mingled with certain of these agents, such as iodine in solution, tannic acid, &c., the *constitutional symptoms declared themselves* as usual in the pigeons employed; but the local phenomena were more or less wanting.

The Crotalus venom has the power of preventing the germination of canary and mignonette seeds.

"On several occasions," I quote the words of Dr. Mitchell, "I had noticed the production of fungi in moist venom, long kept upon my table, in an atmosphere of from 64° to 70° Fahr." "I have also observed in specimens long kept, and somewhat diluted, that after seven to ten days, the poison acquired an odor of a peculiar and very disgusting character. The production of this animalized and indescribable stench, was accompanied by the appearance of vibriones, and a few days later, of rotiferal and other minute forms of animalcular life. The question was, what power yet remained in the venom which had become the nidus of so much active vitality? I tested it on animals, and still found it actively poisonous."

Dr. Mitchell's experiments give the following, as the

*Effect of the Venom upon the Crotalus.**

"This research resolves itself into two propositions, or rather, questions. First, Can the Crotalus kill its own species? Second, Can any individual snake kill itself?

"The first of these queries has been more or less completely answered, as regards certain India snakes, the Viper of

---

* Researches upon the Venom of the Rattlesnake, by S. Weir Mitchell, p. 60.

Europe, and our own Crotalus. Russell* made a Cobra bite a Nooni-Paragoodoo near the anus. It died in one hour and a quarter. A little local discoloration existed about the wound, and the lungs were full of blood. A Cobra bit another Cobra, with a negative result. How long it was observed, is not stated.

"A Coodum-Nagoo bit a Cobra, the two fangs taking effect, the result as before, being negative. All of these are venomous snakes.

"A Coodum-Nagoo bit a Tortulla, a harmless serpent, which perished within two hours.

"Fontana's† experiments on the effect of the venom of the Viper upon its own kind, were briefly as follows:

"One viper was bitten by another several times. The wounds swelled a little. It was killed by Fontana after thirty-six hours, and found to have been deeply wounded, the bites being a little inflamed and swollen.

"A middle-sized viper received from two large ones six fang-wounds. The viper remained agile, and was well at the end of four days. When killed it was found to have been bitten through and through. The wounds were somewhat inflamed. Five other vipers thus bitten did not die. Length of observation not mentioned. Again, a portion of the skin having been removed from the backs of four vipers, seven vipers were made to bite them. None of the bitten animals died, and only one of them was at all languid, and had a little swelling about the wound. Three vipers were wounded in the back, and the wound filled with venom. The wounds inflamed but did not swell. The animals seem to have been killed at the end of several days. A viper was forced to bite

---

* Russell, p. 56.
† Fontana, vol. i, p. 29 *et seq.* Skinner's Translation.

itself; but did not die. Another was made to bite upon a piece of jagged glass so that its mouth was wounded as the poison flowed into it. On the seventh day the wounds were healed.

"M. Bernard* recently repeated Fontana's experiments, and found that a viper which had been both bitten and inoculated artificially with the venom, died on the third day. Upon this experiment M. Bernard criticizes Fontana, as having observed the viper and pigeons together, and having concluded that, because the cold-blooded animal was not so soon affected as the other, it was incapable of being killed by the venom. As we have seen, however, some of Fontana's experiments were observed during periods of time much greater than that required to destroy the viper observed by M. Bernard. Thus, although Fontana was most probably mistaken in his conclusions, he did not fail in the point criticized, from any glaring neglect of continued observation.

"The American authorities upon this matter are brief, but decided. They refer principally to the power of the snake to destroy itself, and to this point, indeed, my own experiments have been directed, since it was plain that if the individual could thus be made to kill itself, there could be no added difficulty in comprehending its ability to kill its fellows.

"Besides including the general proposition, the question before us has a specific interest, from the fact that snakes are often accidentally hurt about the mouth, and that abrasions of this cavity must frequently occur. We are, therefore, called upon to say why the snake suffers so little from wounds on which a poison so deadly to other animals must fall from time to time.

---

* Claude Bernard, Leçons sur les Effets des Substances Toxiques, &c., 1857, p. 291.

"Our own writers* state almost unanimously that the Crotalus is able to kill itself. Without quoting them in full, it is enough to add that their experiments were commonly made by switching a snake until it turned and struck itself. Death is usually described as following within a few moments.

"At the close of a series of experiments on warm-blooded animals, I made use of some of my largest snakes in the following manner:

"*Experiment* No. 1. Temperature 65° to 75° Fahr. A small snake, about twenty-seven inches long, was caught by the neck, and its tail placed in its mouth. It bit, but did not wound. A portion of the skin having been removed from the back, it was allowed to bite again, and when the fangs were fixed in the naked muscles, the upper jaw was violently pressed downwards, so as to wound the part deeply. Upon the sixth day, the wound was covered with a gray exudation such as is usually found upon the healing surface of the wounds of serpents. This snake died on the fourteenth day. The tissues about the bite were congested, the gall-bladder full, mucus in the stomach, the venom-glands dark from effused blood.

"*Experiment* No. 2. A large snake was made to bite himself twice, in a space near the cloaca where the skin had been removed. This serpent also died on the fourteenth day. The wound was apparently healthy, and not to be distinguished from any other wound, except that the muscles about it were a little softened. The blood was uncoagulated, but there was no other visible lesion of any internal organ.

"*Experiment* No. 3. On the same day a large snake, fifty-six inches long, had a small portion of the skin on the back

---

* Burnett, p. 328.

loosened and turned over, so as to make a flap. On this wound was placed about a drop of venom from the snake itself. The poison was finally thrust into a number of superficial cuts made in the muscles on which the drop fell. On the second day, the snake being well to appearance, half a drop of its own venom was put in a superficial wound half way up the back. This wound seemed to excite the snake, which, on being replaced in its box, continued in very rapid and violent motion for some minutes, as though in pain. On the sixth day, both wounds were covered with gray exudation, and beneath, the muscles were soft; but in this, as in other cases, no effusion of blood existed about the wound. The snake was sluggish, and indisposed to bite. It died on the tenth day.

"*Post mortem.* There were no visceral lesions, except that one lung contained a little effused blood. The venom-glands were dark and congested; the heart-blood coagulated firmly thirty minutes after removal. In all probability, this serpent died from some other cause than venom poisoning.

"*Experiment* No. 4. A snake forty-six inches long was secured, and the skin just above the anus removed from a space of about one inch by two. On this the snake bit itself three times, throwing out a good deal of venom, which was thrust deeply into the muscles of the part. On the second day, the wounded muscles were softened, but no blood was effused. The wound had been recovered with skin, and secured by sutures. At the close of two weeks, this snake was healthy, and bit eagerly; the wounds were partially healed.

"*Experiments* 5, 6, and 7. Three large serpents were made to disgorge their venom, and the poison of each snake was injected under the skin of its back with the aid of a small syringe and trocar. The snakes, which I will distinguish as

Nos. 5, 6, and 7, received respectively ten, eight, and seven drops of poison.

"No. 5 died in thirty-six hours. The wound was surrounded by softened tissues, but was not stained with blood. The organs generally were normal, except the stomach, which contained bloody mucus. The heart was full of clotted blood.

"No. 6 died in sixty-seven hours. The local appearances in this case were much as in the last one, but less extensive. The interior organs were healthy, and the heart contained two loose and soft clots.

"No. 7 died during the seventh day. The wound, in this case, penetrated the muscles, which were dark and much softened. The blood in the heart was mostly diffluent, presenting but a single small coagulum of loose structure. The intestines were spotted with ecchymoses, and the peritoneal cavity contained about a drachm of fluid blood.

"I may add to these cases the numerous instances in which I have wounded the mouths of snakes, or torn the vagina dentis, while robbing them of poison. On none of these occasions has any serious result followed the injury, even where venom had fallen upon the abraded surfaces in considerable amount.

"The above experiments were on the whole so definite in their results, that I did not think it necessary to multiply them. I had very many times injured snakes far more than these were injured by their own fangs or the preparatory manipulations, and I therefore felt at liberty to conclude that the animals employed on these latter occasions really died from the venom. The length of time required for this to occur was curious, and far exceeded in most of them that which was noted in Bernard's cases, or in the many instances of which I have been told rattlesnakes had stricken themselves.

"One of the factors in Experiments 5, 6, and 7, and one which has been too much neglected, is the temperature, which in my own cases was very moderate during the day, and fell a good deal lower during the night, the observations having been carried on during a cool period in September, 1859. MM. Bernard, Russell and Fontana give no record of the temperature during their observations. That it is a very important condition in the venom-poisoning of the cold-blooded batrachia, I have frequently observed; and it is highly probable that in all cold-blooded animals the elevation of the temperature carries with it an increase of danger from poisons, and especially from those of a septic nature.

"When we examine the pathological effects of the venom in warm-blooded animals, we shall see that, while the general phenomena were essentially the same as in cold-blooded reptiles and batrachia, they were far more rapidly produced. The Crotalus itself was a good illustration of this contrast, and was in other respects exceptional in the mode in which it was affected; since, while the muscles were altered, as in warm-blooded creatures, the blood coagulated better than was usual in them, and the visceral lesions were less severe and less frequent.

"On the other hand, while the frog for its size is remarkably unimpressible by Crotalus venom, the phenomena which in it accompanied the examples of slow poisoning were in no respect different from those developed in the warm-blooded animals."

From the same work the following interesting data are taken, as they exhibit phenomena having reference to yellow or bilious fevers:

"Both pigeons were enfeebled by the poison, and seemed disposed to sleep; one of them sank slowly lower and lower, until its head touched the table, when it rolled on its side.

"A pigeon, in ten minutes after it was struck on the back by a small snake, fell into the *usual stupor*, with jerking, abrupt respiration, &c.

"The bird became gradually weaker and weaker, and died without convulsions, at the close of an hour and a half. The pupils gradually contracted before death, and suddenly dilated afterwards.

"The tissues, for an inch or more around the wound, were soaked with extravasated blood, which had even passed through between the ribs, so as to stain the tissues behind the intestines.

"The heart was large and full of perfectly fluid blood.

"My chief reason for recording at length these cases," continues Dr. Mitchell, "is to show the *great increase in the internal lesions which occurs, when the venom is long in killing the animal.*"

Among these changes it was found, as a general rule, that the blood was most affected, and least coagulable the longer death was delayed.

In two cases the dogs passed clay-colored stools.

So far as a fatal result is concerned, it seems to be indifferent, whether the bite takes place about the head and neck, or in the limbs.

This last observation is of great interest, owing to the fact that the dog, a small terrier, survived very serious visceral lesions, and lived during two days, with his blood in a condition of complete diffluence.

Twenty hours from the time of the poisoning, the dog was found lying on his left side, having passed slimy and bloody stools in abundance. At intervals he seemed to suffer much from tenesmus, but was so weak that he stood up with difficulty.

His gums were bleeding, a symptom I had seen before, and his eyes were deeply injected. (The most characteristic symptom of yellow fever, according to Manzini.)

Twenty-seven hours and a half after the time he was bitten, his hind legs were twitching, and the dysentery continued. No clot was found in the blood. This case is doubly important, because it shows how long the poisoning must continue before the blood becomes diffluent.

Dr. Mitchell continues:

"The study of envenomed blood has thus far taught us:

"1st. That in animals which survive the poisoning for a time, the blood is so altered as to render the fibrin incoagulable.

"2d. Experiments in and out of the body have given proof that this change is gradual, and that the absence of coagulation is not due to checked formation of fibrin, but to alterations produced by the action of the venom in that fibrin which already exists in the circulating blood.

"3d. The influence thus exerted is of a putrefactive nature, and imitates, *in a few hours,* the ordinary work of days of change. It is probably even more rapid within the body, on account of the higher temperature of the economy.

"4th. The altered blood retains its power to absorb gases, and thus to change its own color.

"5th. The blood-corpuscles are unaffected in acute poisoning by Crotalus venom, and are rarely doubtfully altered in the most prolonged cases which result fatally.

"6th. The contents of the blood-globules of the Guinea pig can be made to crystallize, as is usual after other modes of death.

"Among the most constant and curious lesions in the cases of *secondary poisoning are the ecchymoses, which are found on and in the viscera of the chest and belly, most frequently affecting the intestinal canal.* They may and do occur in any cavity and on any organ. These spots contain blood, whose globules are more or less deformed, but still of dimensions not less than usual.

"Owing to the changes of the blood, or the tissues, or both, extravasations are met with in the lungs, brain, kidneys, serous membranes, intestines, and heart. As a result we may have functional derangement grafted on the main stem of the malady, and the accompaniments of bloody serum in the affected cavities, bloody mucus in the intestinal canal, and bloody urine in the bladder."

From the preceding experiments Dr. Mitchell naturally arrives at the conclusion:

"That the venom of Crotalus, like that of other snakes, is *a septic, or putrefacient poison, of astounding energy;* a view long held by toxicologists.

"*The rapid decomposition of the blood, and of the tissues locally acted upon by the venom, leaves no doubt upon the matter, and makes it apparent that an incipient putrefaction of this nature may so affect the blood as to destroy its powers to clot, and perhaps, also, to nourish the tissues through which it is urged.*"*

The alterations thus brought about are probably the results of a continued fermentative change, which, by a small amount of poison, is gradually made to involve in fatal change the whole mass of the circulating fluids. *Like all fermentations, however, the rapidity depends upon temperature, and on the amount of the primary ferment.*

---

* See Sfigmas, pp. 177–184.

In one instance a dog, struck by eight snakes, died in eighteen minutes, and exhibited uncoagulable blood. I am aware of no other case of loss of uncoagulable blood so rapid.

It was rendered thus by the number of localities from which the ferment attacked the system.

On the other hand, the frog, a small animal, receives the same dose of venom as would have entered the tissues of a larger animal; yet it resists the poison most remarkably, by virtue of its powers as a cold-blooded creature, existing at the temperature of the atmosphere itself.

The cause of death in *chronic or secondary poisoning*, may, with propriety, be referred to the incipient putrefactive changes which affect the blood, as well as the continued influence of the agencies which first act to depress the heart's action, and destroy nerve function.

Summing up what we have learned of the *acute forms of poisoning*, we may feel justified, according to Dr. Mitchell, in concluding :*

1st. That the heart becomes enfeebled shortly after the bite. This is due to *direct influence of the venom on this organ*, and not to the *precedent* loss of the *respiratory function*. Notwithstanding the diminution of the cardial power, the heart is usually in motion, after the lungs cease to act, and its tissues remain for a time locally irritable. The paralysis of the heart is, therefore, not so complete as it is under the influence of *Upas* or *Corroval* poison.

2d. That in warm-blooded animals artificial respiration lengthens the life of the heart, but does not sustain it so long as where the animal has died by *Woordra* or decapitation.

3d. That in frogs the heart-acts continue after respiration has ceased, and sometimes survive until the sensory nerves

---

* Crotalus Horridus, &c., by Dr. Neidhard. New York, 1868. Radde.

and the nerve-centres are dead, the motor nerves alone remaining irritable.

4th. That in warm-blooded animals respiration ceases owing to paralysis of the nerve-centres.

5th. That the *sensory nerves* and the *centres of nerve power* in the *medulla spinalis* and the *medulla oblongata lose their vitality before the efferent or motor nerves become affected.*

6th. That the muscular system retains its irritability in the cold-blooded animals acutely poisoned, for a considerable time after death.

7th. That the effect of the venom being to depress the vital energy of the heart and nerve-centres, a resort to stimulants is clearly indicated as the rational mode of early constitutional treatment.

ANNUAL DEATHS FROM SNAKE-BITES IN INDIA.

According to official returns in the Bengal Presidency for 1869 the following abstracts are taken, viz.:

Bengal, Assam, Orissa,	6,645
Northwest Provinces,	1,995
Punjab,	755
Oude,	1,205
Central Provinces,	606
Central India,	120
	11,416

Making a total of 11,416 for an estimated population of 121,000,000, which would give over 25,000 deaths for the year throughout the whole of India!

Details of the above Cases of Snake-bites, from the preceding Report.

The returns from the Bengal Presidency are detailed as follows, viz.: 3037 males and 3182 females; of these, as far

NATURE AND ANALYSIS OF THE POISONS.

as known, the following table shows which kinds have caused the greatest proportion of deaths in the different districts, viz.:

	Bengal.	Orissa.	Assam.	Oude.	Central India.	Central Provi'ces	N. W. Provi'ces
Cobra,	959	128		607	21		854
Krait, Bung. cœrul.,	160	2		105			92
Other snakes,	348	52	12	20	37		63
Not known,	4752	168	64	473	32	606	986

British Burmah.
Cobra, ... 45
Daboia, . . ⎫
Hamadryad, . ⎬ 75   Other snakes, .
Hydrophis, . ⎭   Not known, ... 437

Punjab.
76
. . . 242
. 437

Dr. Fayrer says,* in all cases where the blood forms a firm coagulum after death the poison is of a *Coluber*, and in all cases where it remains perfectly fluid it is of a *Viper*.

This point is not positively determined as yet. We may, however, take the *Naja tripudians* as heading the scale of those poisons whose action on the blood produces a coagulum, and the *Crotalus* as the synonym for the opposite class, whose action on the blood produces permanent fluidity; it is probable that the action of all the other snake poisons ranges between these two extremes. But under certain conditions the same kind of poison, taken from the same reptile, produces such widely different effects that sometimes it kills quickly, at other times it kills slowly, and during a certain annual period in the life of the reptile it does not produce death, but causes a partial septicæmia only.

To explain, viz.: Just at the time when the snake begins

---

* The Thanatophidia of India, by J. Fayrer, M.D. London: J. & A. Churchill, 1872.

to change its skin, and enters into the state of semi-torpidity consequent upon this change, the poison loses its venomous principle, and what is still more singular *the gall loses its bitter principle and becomes sweet to the taste.*

The poison extracted from the reptile in this state is a milky, thick, viscous fluid, which soon separates itself into a white ropy sediment, and a supernatant, transparent, slightly viscous liquid, showing no trace of acidity. Neither the sediment nor the colorless liquid alone or in combination causes death in pigeons or dogs, but a septicæmia is produced which is limited, and not profound in its action. The poison continues in this condition until the state of torpidity ceases, and when the colors on the new skin are very brightly marked and distinct *the venomous principle is again restored to the poison and the bitter principle to the gall.*

These facts afford us a key to the widely different results obtained by experimenters, who use different substances as antidotes in cases of snake-bites, which appear to be efficacious many times, and not efficacious in many other instances, when really they deserve no merit whatever for efficacy.

Another fact is worthy of notice in this connection: the action of the bile in the snake is precisely the same as the gastric juice in the human stomach. After the reptile has swallowed its food a discharge of the bile into the digestive canal takes place; next ensues a precisely similar condition to that which occurs in jaundice. The whole muscular system becomes saturated with it, and the remaining portion which decomposes the food passes off in the fæces. In this condition if its own or the poison of another snake is introduced into the body, death does not ensue; because the bile which has already been absorbed by the muscular system antidotes the action of the poison, however fully its venomous principle may have been developed when injected.

The poison and the fang by which it injects it are the reptile's weapons of defence, consequently nature does not teach it to use them against its own kind, so it may bite another snake without injecting poison into the wound; whereas, if it bites a man or an animal the injection of the virus follows, an instinct of nature. A knowledge of the foregoing facts leads us to the following conclusions, viz.:

1st. To determine the nature of any of these poisons by experiment we must first be sure that they are taken from a reptile at the period during which the venomous principle is fully developed.

2d. A hypodermic syringe is the proper instrument to be relied on for introducing the virus into the blood, as its action is precisely similar to that of the fang.

3d. We must be sure that there has been no general absorption of bile into the system of the animal to be experimented upon, at the time of making the experiment.

By observing all these conditions, our experiments will be of value to science.

The experiments of Blake, Hering, and Claud Bernard, show that the absorption of the poison takes place with such rapidity as to render the administration of any remedy utterly useless in many cases. The general rule is, that an antidote will prove efficacious in all cases *where no organic lesion has already taken place.*

In South America the bite of a *Vipera Echis variegata* has been known to produce death *in five minutes* in a man; and a *Vipera Lachesis niger* has killed a stout, able-bodied man in as short a period of time.

Blake says* poison passed through the jugular vein to the lungs of a dog in from 4 to 6 seconds; through the jugular

---

* Guy's Forensic Medicine, page 388, third edition.

vein to the coronary arteries of the heart in 7 seconds, and was distributed through the circulation in 9 seconds.

Dr. Fayrer recommends the application of a ligature without delay in cases of snake-bites; but there is not a single case on record where its application has proved of any use whatever, and this fact also accords with my own experience.

To test this point, Dr. Shorlt, of Madras, seized the prominent part of the muscles of a dog's hind leg with a stout pair of forceps, under the points of which a ligature was passed with a strong cord; this was drawn as tightly as possible with the force of a strong man, and the portion of the muscle remaining above the ligature was of the size of a grain of cocoa. An assistant held a pair of scissors just over the ligature. The fang of a live Cobra was passed through the part just above the scissors, and, the instant the fang was drawn, the projecting part was excised close to the ligature.

What appreciable portion of poison could have passed into the circulation in the second or two that elapsed between the injection of the venom and the excision of the part? Yet death ensued in an hour! Had the dog been bitten by the snake in his hinder parts death would have ensued in from 30 to 40 minutes. This was a case in which the deadliness of the poison could not be doubted; but the application of the ligature only delayed the mortal result a few minutes.

To determine the degree of virulence of the different poisons it will be necessary to institute a series of experiments with animals of the same kind, of as nearly the same age as possible. The poison must be introduced into the same vein, *at the same distance from the heart,* and in (as nearly as possible), the same quantity.

In the experiments which have already been made, and in the cases of snake-bites given in another part of this work, the facts prove that, when the poison is introduced into the extremities or the small arteries or veins, death ensues after a much longer time than when it is introduced into one of the larger veins or near the heart; as, in the latter case, death ensues almost invariably within a few seconds. This fact is particularly noticeable in the action of the *Naja trip.* poison.

The following experiments were made in St. Bartholomew's Hospital, in London, in November, 1872, in the presence of the following gentlemen: Dr. J. Forbes Watson, of the India Office; J. Fayrer, M.D., author of the Thanatophidia; Dr. T. Lauder Brunton, casualty physician to the hospital, and the author. The Cobra poison used in the experiments was very much decomposed, with its accompanying unbearable stench, having been kept in the laboratory for more than a year. It was diluted in the proportion of about one drop of poison to twenty drops of water.

Experiment No. 1. With a tame pigeon, a dilution of the Cobra poison, injected into the left hind leg, at 1.46½ P.M. At 1.50 it was seated with the point of its beak resting on the floor; did not appear to be in pain. At 1.52, or in 5½ minutes, dead.

Experiment No. 2. Another pigeon, of the same size and age, had injected into its left hind leg a dilution of the same poison at 1.56 P.M. At 1.59 a dose of the antidote, as prepared in Experiment No. 4, was injected into the mouth, by means of a syringe; at 1.60 dose of antidote; at 1.62 dose of antidote; at 1.62½ wound moistened with antidote; at 2.4 dose of antidote; at 2.7 dose of antidote; at 2.15 dose of antidote; at 2.21½ P.M., or in 25½ minutes, dead.

Deductions from the previous Experiment. Although the antidote was not prepared from the gall of the Cobra, yet, still it seemed, to a certain degree, to counteract the effects of the poison. In the preceding experiments, the point of the syringe, as far as could be judged, did not penetrate any principal artery or vein.

Experiment No. 3. With a white rabbit, apparently in perfect health, and of a medium size.

At 2.51 P.M., injected 20 minims of the liquid into the femoral artery of the left hind leg.

At 3.33 P.M., nearly dead.

At 3.41 P.M., dead.

At 3.42 P.M. an opening in the larynx was connected with a small pair of bellows, by means of an elastic tube, and artificial respiration was commenced; the heart commenced beating at the rate of 150 pulsations per minute, which continued till 5.15 P.M., when the respiration was discontinued, owing to the lateness of the hour.

Death ensued in this case in 50 minutes after the injection of the poison.

An experiment had been made by Dr. Brunton a few days previously, in which a rabbit had been killed by the Curare poison, and in which artificial respiration was sustained for four hours and a half, when the animal was restored to life, the poison having been eliminated in the involuntary discharges of fecal matter and urine, which continued during the time of artificial respiration.

Two days later than Experiment No. 3, another experiment was made of a precisely similar nature, in which artificial respiration was kept up for six hours, when the pulsations of the heart had diminished from 150 to less than 50, and as it was evident that they must soon cease entirely, the artificial respiration was discontinued.

Deduction from Experiment No. 3. 1st. That the sensory nerves were first affected by the poison, the source of nerve-force having been apparently annihilated by it.

Experiment No. 4. An albino rabbit, of the same size and age as the first one, had injected into the femoral artery of the left hind leg twenty minims of the diluted poison at 2.53½ P.M. It was remarked in this case that the point of the syringe penetrated the artery. Made a preparation of five drops of the first decimal dilution of the gall of a South American Viperine snake in four ounces of fresh water. Gave of this, at 2.59 P.M., first dose; 3.03½ second dose; 3.08 P.M. third dose; when it became evident that the antidote was producing no effect, and it was discontinued. At 3.28 P.M., or *in thirty-four and a half minutes*, death ensued. At 4 P.M. a portion of the blood, poured into a deep dish, had formed a firm coagulum, with very little apparent change of color.

Deductions from the foregoing Experiment. That as the point of the syringe had penetrated the artery, and as the albino organization is so peculiarly sensitive in every respect, death had ensued much quicker than in Experiment No. 1. As to the action of the antidote, it was evident that the gall of the Cobra was necessary to counteract the effects of the Cobra poison, as the range of action of the gall must necessarily be similar, in a pathogenetic (therapeutic) sense, to the action of the poison in a toxicological sense.

Experiment No. 5. With the poison of a South American Viperine snake called *Bogui Dorada* (or Gilded Mouth)—a quantity of about four drops of the poison to each ten minims of liquid.

At 3.14½ injected twelve minims near the femoral artery of the right hind leg of a black rabbit, of same size, age, and weight as No. 4. At 3.19½ tried to force into his mouth a dose of same antidote as used in Experiment No. 2, but he obstinately refused to swallow. 3.27 P.M. tried to give a dose of antidote—same result; 3.44, ditto; 3.53, ditto. 40.3, at Dr. Fayrer's suggestion, injected fifteen minims more of the diluted poison in the same leg. At 4.16 tried unsuccessfully to make him take a dose of the antidote. At 5.15 P.M. the poison had not taken any apparent effect.

Experiment No. 6. At 3.17 P.M. injected into the hind leg of a gray rabbit, of the same size, weight, and age as No. 5, twelve minims of the same dilution of poison. As no apparent effect had been produced at 4.04½, injected ten minims more of the diluted poison. At 5.15 P.M. no apparent effect had been produced by the poison; but about thirty-six hours thereafter, rabbit No. 6 died, having had continuous evacuations for the previous twelve hours. Forty-eight hours after the injection of the poison, rabbit No. 5 died, without the evacuations.

Experiment No. 7. A white rabbit, of the same size, age, and weight as the preceding, had one drop of the Cobra poison of Experiment No. 1, diluted in twenty drops of water, injected at 4.35 P.M. into the jugular vein. As soon as the syringe could be detached from its nozzle, another syringe was attached to it, containing a solution of three parts water and one part liquid ammonia of 959 specific gravity, which was injected instantly into the vein. Dead in fifty seconds.

Experiment No. 8. Same poison; rabbit of same age, weight, and size. Ten drops of the liquid injected into the jugular vein at 4.42 P.M. An injection of forty minims of the same liquid ammonia diluted with water was forced into the vein, with a delay of not exceeding five seconds. At 4.43 dead.

The rabbit No. 3 had its intestinal cavity opened at 5 P.M. The peristaltic motion of the intestines had not been annihilated; the blood formed a firm clot in ten minutes after exposure to the atmosphere.

Deductions from the previous Experiment. There can be no possible doubt that the Cobra poison kills the source of nerve force, and that the reflex action through the motor nerves is remotely influenced by the action of the poison.

Professor Halford* cites twenty cases of snake-bites in which hypodermic injections of liquor ammonia were used, of which seventeen cases recovered and three died. The quantity injected varied from thirty-six minims of liquor ammonia (B. P.), 959 specific gravity, to 30 minims + 30 minims water; also 10 minims liquor ammonia fortissimus + 50 minims water; also 5 minims liquor ammonia fortissimus + 25 minims warm water,—varied in different cases according to the age, sex, and apparent strength of the patient.

These cases were all of such a nature (I quote the Professor's own words), that those who had charge of the patients believed that death would have ensued unless the liquor ammonia had been used.

Of the effect of ammonia on the circulation, Professor Halford continues: "It may be stated as a rule with but few exceptions, in cases of snake-bites, that the blood loses its power of coagulation, and becomes thinner and poorer, its color resembling the most deodorized form of Stokes's cruorine.

It greedily absorbs oxygen, however, after death when exposed to the air, and it absorbs it with greater readiness than unpoisoned blood. According to Dr. Harley's observations in a dog poisoned by a Puff Adder, the blood absorbed two per cent. more oxygen than healthy blood similarly treated.

---

* In a paper read before the Victoria Society of Melbourne, Australia, by Dr. George B. Halford, June 1, 1870.

Why, when no impediment exists to the access of oxygen, should the blood become so gradually deoxidized in cases of snake-bites, except upon the supposition that something had been added to it which prevented those actions and reactions between blood and oxygen upon which the manifestations of life depend? The fluid state of the blood is also in harmony with deficient oxidation, and yet there is no hindrance to respiration, no reason, as in drowning, why the oxygen should not reach the blood. It does reach it, but in the presence of serpent-poison its life-sustaining power is withheld, unless some other agent be added to the blood to counteract it. When the poison is in the blood, the large cells are strangely altered: the white corpuscles are greatly swollen, and certainly increased in quantity, the increase in size apparently keeping pace with the fluidity of the blood, although such changes do not occur in white corpuscles placed in water. This change in the corpuscles commences during life, but goes on indefinitely after death, until all the granular matter seen in the blood becomes converted into cells.

Their average diameter is $\frac{1}{1700}$ of an inch. The nucleus is round or kidney-shaped, and the outer cell-wall so delicate as to escape the notice of most observers until their attention is repeatedly directed to it. In addition, the application of magenta reveals a minute colored spot, like a ruby, at some part of the circumference, resembling the macula seen in the red corpuscles of all vertebrates, as pointed out by Roberts.

This condition of white cell is common to nearly all fluid blood of cats and dogs; especially where it is deficient in fibrin the cells become very evident after death, but when the blood is rich in fibrin the corpuscles, being less at liberty, retain their normal shapes; in fact they swell in serum minus fibrin. In leucocythæmia, where both white cells and fibrin are in excess, these changes do not occur.

These changes in the blood I refer to germinal matter introduced from the serpent's glands, and all the symptoms to such changes. This germinal matter consists of nuclei $\frac{1}{4000}$ of an inch in diameter; the fluid being either slightly acid or neutral. To explain: there is little if any difference chemically between starch and sugar, but physically they are very different; little difference chemically between fibrin and albumen, physically they are very different: but both starch and fibrin are rendered soluble by the addition of minute particles of germinal matter; in the first case, from the salivary glands; in the second case, from the stomach. True, with the latter is combined an acid, but without the pepsin-cells no digestion takes place, and what is very remarkable, a very minute quantity will convert a very large quantity of fibrin into albumen so long as the mixture remains acid; the pepsin-cells require no renewal, as they are said to determine the conversion by catalysis.

A little spittle from a man's mouth will change starch into sugar in one minute. This being the fact, it is unreasonable to claim for another gland closely allied to the salivary, an equal power in another direction.

What the changes produced in the blood are, we do not fully know; but that which in healthy blood is called fibrin is diminished, digested, if you will, in the presence of serpent's poison. If the quantity of this substance entering the circulation is great, the changes are greater, and chances of recovery less. If the animal is small its chances of recovery are still less; if it depends much upon oxygen, with a naturally high fibrinous blood, rapid circulation and absorption, it soon falls a victim; thus a bitten bird succumbs in a few minutes.

If the animal be in a degree indifferent to oxygen, *i.e.*, can live a long time under water, in carbonic acid gas, as a reptile or hibernating animal, with correspondingly slow circulation and absorption, its chances of recovery are greater; it is naturally more indifferent to the presence of the poison, and the probability of absorption from a wound is less.

It may be said that the symptoms of snake-poisoning come on too soon for such changes in the blood to be produced; but recollect the nearly instantaneous action of pytalin. If venom enter the vein, one minute will certainly suffice for its entire circulation through the body.

Estimating the quantity of blood in the body at twelve pounds, and the contents of the left ventricle at three ounces, then the whole mass of the blood would be circulated by sixty-four heart-beats; and taking the normal beats at seventy-five per minute, this quantity of blood would pass through the heart, carrying with it the imported poison in fifty-one seconds.

I have shown by experiments that dyes, such as magenta, may be absorbed from the serous cavities, peritoneum and pleura, and from the subcutaneous connective tissue, and traverse the circulation, be excreted by the kidneys, and ejected by the bladder, in less than five minutes. The experiments proving this fact were made upon dogs and fowls.

Later, Drs. Macnamara and Haughton have found iodine pass from the tunica vaginalis testis, and to be discharged from the bladder in four minutes; therefore in man, symptoms of poisoning may certainly be possible in five minutes, and in birds in half that time.

Believing then, that in the blood itself the chemical changes occur which have their resultants in voluntary and reflex acts (will and motion), and in animal heat, let us turn to the condition of our snake-bitten patient. He is usually pale, with a great tendency to sleep, the heart's action is feeble, and in some cases a deep coma ensues, from which it is difficult to arouse him; in fact the countenance is like that of an epileptic after the convulsion.

Do not the pallor and drowsiness indicate an anæmic state of the brain, or rather a diminution of the arterial capacity and extension of the venous surface analogous to that which takes place in a less degree in ordinary sleep, according to the researches of Dr. Durham? In all the cases of cures cited, the functions of the cerebral ganglia and higher sensorium were for a time extinct.

The sufferers were dead to both sound and light, and had absolutely no knowledge of the injections to which they were submitted.

Now it is worthy of notice, that recent views ascribe the phenomenon of sleep to the using up of the previously stored-up oxygen in the blood, which is analogous to the accession of sleep in cases of snake-poisoning.

Another symptom which deserves the greatest consideration is the extreme dilatation of the pupil, depending, as I believe, on the central optic ganglia.

Light is pouring into the eyeball, and yet no reflex acts are produced. An exactly similar condition affects the auditory ganglia; vibration after vibration shocks the tympanum, each falling silently upon the central ganglia. Sometimes this condition of the sensory ganglia, including probably the corpora striata, optic thalami, and corpora quadrigemina, exists, while the higher cerebral ganglia either remain a long time unaffected or escape entirely, in the latter case, recovery resulting; in the former, constituting a most dangerous state, especially when accompanied by swelling or pain in the neck.

Mr. Hodgkinson, according to his own experience, after being bitten by a snake, says: "There was insensibility to the effects of pungent salts when held to the nostrils, and to pinches inflicted by some of the lookers-on, who desired thereby to prevent me from falling into a comatose condition. There was also deprivation of sight, which attained its maximum about two hours after being bitten."

Mr. Hodgkinson we all know is no ordinary man, and I have no doubt that his intellectual activity or will, contributed to save him.

Professor Halford then recommends injecting dilute ammonia, an old and favorite remedy, he says, the world over; but not to be put into the stomach, the stagnating vessels of which cannot absorb, but injected into the blood itself.

On the latter point, viz., that the vessels of the stomach cannot absorb quickly, almost every homœopathic practitioner, who has had even a limited practice, would join issue with the Professor; for the facts are, that sometimes medicines under certain conditions, do produce the most extraordinary results in a few seconds in some cases, in a few minutes in others; by being introduced into the stomach, and it is quite evident that such results could only be produced by absorption.

The action of preparations of the gall taken into the stomach produces effects precisely similar to the preceding in cases of snake-bites; whether these preparations would be more efficacious in hypodermic injections I cannot say, but a series of experiments is being carried out in India at the present time (January, 1873), to test this point with Cobra poison. We must also remember that different snake-poisons develop different symptoms.

The Yellow-mouthed Viper (Vipera Lachesis os flavus) of South America, producing, for example, suffusion of blood to the face and head; veins of conjunctiva distended; patient complains of heat, of great pain in the bitten part, great thirst; eyes have a wild look, are very sensible to the light; in many cases the sight is partially destroyed, and there is a discharge of blood from the mouth and nose, from the urinary canal, and at times from under the nails of the extremities.

In fact, from what symptoms are developed in cases of poison of one kind, you arrive at one conclusion by a course of reasoning as to its action, only to have this completely set aside by the effects produced by snake-poison of another kind in other cases. A third kind produces quite different effects, so as to make you question the previously formed opinion; and lastly, you are left in a maze of doubt as to what is true and what not true about its action, by noticing carefully the symptoms developed in a fourth case by still another variety of the serpent's poison.

From the above facts one can readily comprehend some of the difficulties that present themselves to the student in this branch of science, which by the way is so thankless in results that it has occupied the attention of the English surgeons in India for one hundred years past; and they frankly confess* that very little more is known about them than what Dr. Patrick Russell† learned by his studies.

Since 1860, however, our knowledge on this subject has taken a giant stride. Let us hope the day is not far distant when this shall have become exhaustive.

The following article from Fayrer's Thanatophidia‡ is a curiosity in medical literature, because the Oriental doctors of medicine have always shown a disinclination to furnish outside barbarians with any insight into their preparations of remedies or methods of treatment.

---

* See an editorial in the London Lancet, published during the year 1870.

† Transactions of the Royal Asiatic Society for 1780, from a paper read by Dr. Patrick Russell.

‡ Lib. cit.

The Use of Snake-poison in Medicine by the Kabirájes* of Bengal.

Furnished by Babóo Gúnga Pérsed, Senior, one of the most learned Kabirájes in Calcutta.

The poison generally used is taken from the *Keautiah* variety of the Cobras, and not from the *Gokurrah* (spectacled Cobra), *Sunkerchor*, *Sankni*, or *Bora*, as the poisons of these latter varieties are extremely acute.

The snake is introduced into an earthen pot, in which are placed two or three green plantains, the opening of the pot being covered with an earthenware plate. The application of heat to the bottom of the vessel causes the snake to become furious, and he bites the plantains.

The bitten part turns black, and is cut out, dried, and reduced to powder, then purified, and it is ready for use. The poison thus obtained is impregnated with saliva and other impurities, and is said to be in this state an irritant, warm, sharp, penetrating, exceedingly quick in its action, and produces derangement of the nervous and digestive systems; hence learned practitioners of ancient times used to purify it by mixing therewith the juice of *neem* leaves and lime-juice, and then it was dried.

This process is repeated *five times*. They consider its physiological action to be warm, irritant, stimulating; a promoter of the virtues of other medicines; antispasmodic; a promoter of the action of the organs of secretion.

Its Therapeutical Action.—Used in the latter stage of low forms of fever, when other remedies fail, it accelerates the heart's action, and thus diffuses warmth over the whole surface, and dissipates coma. It is used successfully in the collapsed stage of cholera; also in dysentery, and some complicated diseases;

---

* "Fathers in medicine."

in epilepsy arising from cold, it relieves the patient of insensibility and forgetfulness, two well-marked symptoms developed by this disease.

Some practitioners state that it is used in cases of snake-poisoning, where the body is cold and the heart's action hardly sensible. In these cases its use is said to produce a flow of blood to the distant capillaries in which circulation had ceased, thus diffusing warmth over the whole surface. Subsequently antidotes are used which circulating with the blood are diffused over the whole system. Antidotes, unless mixed with the poison, cannot be introduced into the system by reason of the cessation of the circulation. *Moreover, snake-poison is the only medicine that can produce instantaneous effects on the whole system;* for this reason also the antidotes are mixed with the poison. The Kabiráje says, he believes certain vegetable and mineral poisons are rather proper antidotes to snake-poisons, and *vice versâ,* as the latter cause determination of blood to the brain, and thereby affect the nervous system; whereas the vegetable and mineral substances used as antidotes, mostly cause determination of blood to the alimentary canal, and thereby change the position of the congestion from the brain to the alimentary canal.

The *Bish Badis** in general use among the Kabirájes are of three kinds, and are called:

    No. 1. Súchikábarana.
        2. Aghora-nrisinha-rasa.
        3. Pratápa-lankeshvara.

No. 1 is prepared in two ways.—*First method:* A sort of black sulphuretted mercury, 2 parts; burnt oxidized lead, 1 part; aconite, 1 part; snake-poison, 1 part. These ingredients, all reduced to a powder, mixed, and subjected to a process in

---
* Mixtures of antidotes and poisons.

which the powder is repeatedly mixed and pounded with certain liquids, exposed to the sun, and then dried; *i. e.*, the preparation is to be treated with fish gall, 7 times; goat's gall, 7 times; peacock's gall, 7 times; wild boar's gall, 7 times.

A *second method* is: Mercury, 1 part; snake-poison, 2 parts; to be mixed and evaporated in a retort, the vapor collected in a receiver, and kept for use in stoppered vials. This to be used hypodermically at the centre of the scalp in epilepsy and snake-bites.

No. 2. *Aghora-nrisinha-rasa:* Sulphuretted mercury, 2 parts; oxidized burnt copper, 1 part; oxidized burnt iron, 2 parts; oxidized burnt tin, 3 parts; burnt mica, 4 parts; purified red arsenic, 1 part; a preparation of gold and mercury with sulphur, 1 part; snake-poison, 4 parts; ginger (long and black pepper), 4 parts; aconite, 46 parts. These ingredients pounded and mixed together are subjected to the liquid process with the following substances: Fish bile, 7 times; buffalo's bile, 7 times; Rakta-chitra, 7 times; goat's bile, 7 times; peacock's bile, 7 times; boar's bile, 7 times.

Third method, or Pratápa-lankeshvara, is compounded of sulphuretted mercury, 2 parts; burnt mica, 1 part; vermilion, 1 part; aconite, 1 part; borax, 1 part; burnt copper, 1 part; burnt iron, 1 part; burnt tin, 1 part; licorice, 1 part; root and stem of the Sida cordifolia, 1 part; tubers of the Cyperus hexastachyus, 1 part; *Renuká* (an aromatic seed like pepper), 1 part; resin of the Balsamodendron agalocha, 1 part; red arsenic, 1 part; snake-poison, 1 part; flowers of the Mesua ferrea, 1 part. These to be reduced to a fine powder and mixed, and then to be subjected to the liquid process with the following substances, viz.: Decoction of the three pungents, 7 times; juice of Datura stram., 7 times; juice of Cannab. Indica, 7 times; juice of the Rakta-chitrá,

7 times; juice of the Iválámukhí, 7 times; fish bile, 7 times; goat's bile, 7 times; boar's bile, 7 times; buffalo's bile, 7 times; peacock's bile, 7 times.

What is most remarkable in the preparation of these Bish Badis is their final treatment with the bile of animals. Any preparation thus treated has nothing left of the taste of the other ingredients *but the bitter of the gall;* and the Kabirájes must have found in the latter a principle which produced a marked action in antidoting the snake-poisons. This fact argues the existence of elements common to all galls, probably *Cholic* and *Choleic* acids and *Cholesterin*.* The formulæ are: Cholic acid $= C_{52}H_{43}NO_{12}$. Choleic acid, containing a large amount of sulphur, hitherto not obtained pure, but yielding when boiled in alkaline solutions a neutral sulphuretted substance, Taurin $=C_4H_7NS_2O_6$, of remarkakly beautiful crystalline forms.

Cholesterin crystallizes in brilliant, colorless lamellæ, a neutral, insipid, and inodorous substance, slightly soluble in cold, and very soluble in boiling alcohol. It melts at 278.6° Fahr., being decomposed only at a very high temperature; and it resists the action of alkaline lixiviæ. Its true composition has probably never been determined, but it corresponds nearly to $C_{26}H_{22}O$.

CASES OF SNAKE-BITES.

The following cases of snake-bites are from the official returns to the India Government, and published by Dr. Fayrer:†

No. 1. A woman, æt. 36; bitten by a snake; kind not known; died in 28 hours. She was treated by incantations, charms, and native drugs.

No. 2. A man, æt. 58, bitten probably by a Cobra; died

---

* Regnault's Chem., II, p. 746. † See Thanatophidia.

in 1 hour 30 minutes. Received no treatment; complained of much pain in the wound; rigor mortis occurred 8 hours after death.

No. 3. Man, æt. 32; kind not known; died in 42 hours; treatment same as No. 1. Autopsy showed a semi-fluid blood containing clots.

No. 4. Boy, æt. 12; kind not known; died in 18 hours; treated by the Kabirájes. Autopsy showed fluid blood.

No. 5. A man, æt. 46; kind unknown; probably a Bungarus c., or Krait; dead in 10 hours; treated same as No. 1. Autopsy gave fluid blood.

No. 6. A boy, æt. 8; kind unknown; dead in 24 hours; treatment same as No. 1. Autopsy gave fluid blood, heart entirely empty.

No. 7. A man, æt. 50; bitten by a Cobra; died in 2 hours.

No. 8. A woman, æt. 30; kind unknown; dead in 10 hours; treatment same as No. 1.

No. 9. A boy, æt. 12; bitten by a Cobra; dead in 30 minutes; treatment same as No. 1.

No. 10. A man, æt. 30; kind unknown; died at the expiration of 6 days; treatment, hypodermic injections of liquid ammonia. Autopsy revealed fluid blood; no rigor mortis ensued.

No. 11. A man, æt. 22; supposed to have been bitten by a Krait; died in 8 hours; treatment same as No. 1. Autopsy showed the heart filled with fluid blood.

No. 12. A woman, with a child at breast; the mother died in 4 hours; child died in two hours, supposed to have been poisoned by the mother's milk; treatment same as No. 1. The faces of both were livid and swollen, and a bloody froth issued from their mouths and nostrils.

No. 13. A man, æt. 36; kind unknown; bitten on one foot; was taken into the Jainser Dispensary, and treated by a strong caustic lotion and charcoal poultice applied to the bitten part; cured in 22 days.

No. 14. A young man, æt. 18; bitten by a Chokoriah Borah; died in 26 hours, after vomiting a blackish fluid; the snake was supposed to have been a Crotalus.

No. 15. A boy, æt. 10; kind not known; died in 24 hours; was treated by the Ozahs (snake-charmers).

No. 16. A boy, æt. 8; bitten by a Chokoriah Borah in the foot; died in 31 hours; treated by the Ozahs.

No. 17. A man, æt. 35; snake was probably a Trimeresurus; was ill for a day and recovered.

No. 18. A young man, æt. 18; bitten by a Chokoriah Borah; died in 26 hours; treated by the Ozahs; vomited a black matter.

No. 19. A boy, æt. 16; kind of snake probably a Daboia; died in 2 days.

No. 20. A girl, æt. 7; bitten by a Cobra; died in 15 hours 30 minutes.

No. 21. A man, æt. 23; bitten by a Daboia; died in 27 hours; the wound was scarified, and ammonia administered at first; afterwards an opiate was given, and, after 24 hours had elapsed, galvanic shocks were applied.* An autopsy showed the heart to contain semi-fluid blood and clots; the spleen was soft and easily broken up.

No. 22. A young man, æt. 19; probably a Trimeresurus; was well in 36 hours. Treatment: a ligature, amm. aromat., tinct. opii; the bitten part was covered with ipecac. pulv. and ammonia, and 5 minims liquor arsenicalis administered every 6 hours. He was apparently well 3 hours after the bite.

---

* Their application was made too long after the injection of the venom. Why not have tried them at first?

No. 23. A woman, æt. 37; bitten by a Bungarus fasciatus; was well in 24 hours. Treatment: liq. amm., internally; externally, ipecac., amm., and chloroform.

No. 24. Captain S.; bitten by a Gyal or Hydrophis; died in 71 hours. Treatment: internal doses of sulphuric ether, tincture hyosc., camph., pulv. jalap, calomel, tincture Cannabis Ind., chloroform, amm. aromat., camph. Again: pulv. ipecac., amm., hydrate of morphine and water; brandy having been administered freely during the whole course of treatment.

No. 25. A sailor; bitten by a Hydrophis, 7 feet 6 inches in length; died in less than 4 hours. Treatment: external applications of liq. amm., tinct. opii, and olive oil. He could take nothing internally without vomiting it immediately.

No. 26. A female, æt. 26; bitten by a Krait, of the variety called Chŭtah; died in 3 hours 30 minutes.

No. 27. A female, æt. 50; kind of snake not known; no result; was treated by administrations of liq. amm.

No. 28. A sepoy; kind not known; recovered in 13 days; was administered tinct. ferr. mur. Hæmaturia supervened, but ceased after the lapse of 10 days.

No. 29. A young man; kind of snake not known; well after 10 days. The bitten part was touched with nitrate of silver; a snake-stone applied to the wound would not stick. On the fifth day an abundant hemorrhage supervened, which ceased of itself.

No. 30. A man; bitten by a Bungarus cœr.; died in 63 hours. Treatment: repeated doses of ammonia and brandy.

No. 31. A man; bitten by a Korite or Karite; died in 59 hours. Treatment: internal, ammonia and brandy; external, applications of ammonia.

No. 32. A man; kind of snake not known; died in 59 hours. Treatment same as No. 31.

No. 33. A boy, æt. 9; kind not known; died in 15 minutes. No treatment. Forty drops of blood from this patient, injected into the limb of a fowl, produced no apparent effect.

No. 34. A man, æt. 45; bitten by a Cobra; died in 15 minutes. Rigor mortis in 1 hour after death. No treatment.

No. 35. A man, æt. 32; kind of snake not known, but supposed to be a Cobra; died in 3 hours; treated by an Ozah.

No. 36. A boy, æt. 15; snake was probably a Cobra; died in 10 hours. Treatment same as No. 1.

No. 37. A young man, æt. 20; snake was probably a Cobra; died in 1 hour 30 minutes. Treatment same as No. 1.

No. 38. A woman; kind of snake not known; died in about 3 hours. No treatment. Lungs intensely congested; no rigor mortis.

No. 39. A woman; bitten by an adder, kind not known; died same day. No treatment.

No. 40. A man, æt. 30; bitten by a supposed Cobra or Krait; died in 14 hours. Treated with the usual charms and incantations.

No. 41. A man, æt. 30; snake supposed to have been a Krait; died in 12 hours.

No. 42. A man, æt. 27; bitten by a Krait; died in 9 hours. An autopsy showed coagulated blood in the body.

Nos. 43, 44, 45. Three men were forced to allow themselves to be bitten by two Kraits and a Cobra; one died in a few hours, and the other two in the following day. Another man, bitten at the same time, survived. The scoundrels who caused these men's death were sentenced to five years' imprisonment.

No. 46. A lad, æt. 14; snake was probably a Bungarus cœr.; died several hours thereafter.

No. 47. A young man, æt. 18; bitten by a Cobra; died in 3 hours. Treated by the Kabirájes.

No. 48. A man; bitten by a Cobra; well in 2 days; the wound was incised with a knife, and cauterized with strong nitric acid. He was dosed freely with brandy. The snake was caught, and made to bite a fowl, which died in $4\frac{1}{2}$ minutes.

No. 49. A female, æt. 32; kind of snake not known; died in 3 hours.

No. 50. A man, æt. 27; kind of snake not known; died in 34 hours.

No. 51. A woman, æt. 50; bitten by a Cobra; died in 3 hours. Treatment same as No. 1.

No. 52. A man, æt. 30; bitten by a Chundra Bora or Daboia; died in 7 hours. Treatment same as No. 1.

No. 53. A man, æt. 55; bitten by a Bora; died in 9 days.

No. 54. A man, æt. 50; snake said to have been a Cobra; died in 17 hours. Treatment same as No. 1.

No. 55. A man, æt. 60; kind of snake unknown; died in 30 minutes.

No. 56. A boy, æt. 3; bitten by a Cobra; died in 1 hour.

No. 57. A girl, æt. 15; bitten by a Cobra; died in 3 hours.

No. 58. A woman, æt. 27; bitten by a Cobra; died in 1 hour 30 minutes. Treated by charms and incantations.

No. 59. A man, æt. 35; kind of snake not known; died in 24 hours.

No. 60. A woman, æt. 26; bitten by a Cobra; died very shortly after.

No. 61. A man; kind of snake not known; died in 2 hours. No treatment.

No. 62. A man, æt. 35; bitten by a Cobra; died in 1 hour. No treatment.

No. 63. A man; kind of snake not known; died in 8 hours. Treated by the Kabirájes.

No. 64. A girl, æt. 12; kind of snake not known; died in less than 2 hours.

SUMMARY OF CASES. 159

No. 65. A man, æt. 33; bitten by a Bungarus cœr.; died in 6 hours. Treated by the natives.

A greater number of cases of bites could be given, but they are only repetitions of those already noted.

Summary of the Preceding Cases of Bites.

Sex.	Age.	Variety.	Days.	Hours.	Minutes.	Sex.	Age.	Variety.	Days.	Hours.	Minutes.
1. Woman,	36	Not known (Dab.),	...	28	...	Man,	...	Hydr.	...	4	...
Man,	58	Cobra,	...	1	30	Woman,	26	Krait,	...	3	30
"	32	Not known,	...	42	...	"	50	Not known, [no result.			
Boy,	12	"	...	18	...	Man,		" [well in 13 days.			
5. Man,	46	" (Kr.),	...	10	...	Young man,	...	" "	10 days.		
Boy,	8	" (Dab.),	...	24	...	" "	...	Bung. c. (Kr.),	...	63	...
Man,	50	Cobra,	...	2	...	" "	...	Karite (Kr.),	...	59	...
Woman,	30	Not known (Kr.),	...	10	...	" "	...	Not known (Kr.),	...	59	...
Boy,	12	Cobra,	...	...	30	Boy,	9	Sup. Cobra,	...	...	15
10. Man,	30	Not known,	6	...	...	Man,	45	Cobra,	...	...	15
"	22	Krait,	...	8	...	"	32	Sup. Cobra,	...	3	...
Woman,	...	Not known (Kr.),	...	4	...	Boy,	15	Cobra,	...	10	...
Her child,	...	" "	...	2	...	Young man,	20	"	...	1	30
Man,	36	" (cured),	22	...	...	Woman,	...	Sup. Cobra,	...	3	...
Young man,	18	Ch. Bor.,	...	26	...	"	...	Adder,	...	12	...
Boy,	10	Sup. (Dab.)	...	24	...	Man,	30	Sup. Krait,	...	14	...
"	8	Ch. Bor.,	...	31	...	"	30	Krait,	...	12	...
Man,	35	Trim. (Dab.),	recov'd			"	27	"	...	9	...
Young man,	18	Ch. Bor. (su'd),	...	26	...	"	...	Cobra, [A few hours.			
Boy,	16	Sup. Dab.,	2	...	...	2. Men,	...	Krait,	...	24	...
Girl,	7	Cobra,	...	15	30	Boy,	14	Bung. c. (Kr.),	sev'l hrs		
Man,	23	Dab.	...	27	...	Young man,	18	Cobra,	...	3	...
Young man,	19	Trim. (Dab.),	...	36	...	Man,	...	" [well in 2 days.			
Woman,	37	Bung. f., [well in 24 hrs.				Woman,	32	Sup. cobra,	...	3	...
Man,	...	Hydr.,	...	71	...	Man,	27	Not known,	...	34	...
Woman,	50	Cobra,	...	3	...	"	35	Sup. Dab.,	...	24	...
Man,	30	Dab.,	...	7	...	Woman,	26	Cobra,	...	...	45
"	50	"	9	...	...	Man,	...	Sup. Cobra,	...	2	...
"	50	Cobra,	...	17	...	"	35	Cobra,	...	1	...
"	60	Supd. Cobra.,	...	...	30	"	...	Sup. Kr.,	...	8	...
Boy,	3	Cobra,	...	1	...	Girl,	12	Sup. Cobra,	...	2	...
Girl,	15	"	...	3	...	Man,	33	Bung. c. (Kr.),	...	6	...
Woman,	27	"	...	1	30						

Classifying these, gives by Cobra-bites 24 cases, as follows:

2 cases of	15 minutes each.		5 cases of	3 hours each
2 "	30 " "		1 case of a few hours.	
1 "	45 "		1 "	10 hours.
2 "	1 hour each.		1 "	15 hours 30 minutes.
3 "	1 hour 30 min. each.		1 "	17 hours.
3 "	2 hours each.		1 case recovered.	

In 18 cases, death ensued in from 15 minutes to 3 hours; and in the remaining cases, from a few hours (5 or 6) to 17 hours ensued.

The short time in some cases (marked not known) seems to refer the deaths to Cobra poison, and they are classified accordingly.

By Krait-bites there are 16 cases, divided as follows, viz.:

1 case of 2 hours.			1 case of 6 hours.		
1 "	3 hours 30 minutes.		2 "	8 " each.	
1 "	4 "		1 "	9 "	
2 "	10 " each.		1 "	several hours.	
1 "	12 "		1 "	24 hours.	
1 "	14 "		2 "	59 " each.	
1 "	63 "				

Nine cases varied from 2 to 10 hours. Six cases varied from 12 to 63 hours. In one case death ensued in a few hours; and it is remarkable that not a single case known to have been a Krait is reported as having recovered.

The most remarkable case is that of the child of the woman in Case 12, which died in two hours from the poison in the milk sucked from its mother's breast; while the mother did not succumb till 4 hours had elapsed after the injection of the poison.

But one case of Bungarus fasciatus bite is recorded, and in which the patient is reported to have recovered.

Bites by the Hydrophis are recorded in two cases, in one

of which death ensued in 4 hours, and in another in 71 hours; the latter case having undoubtedly been prolonged by the constant and repeated administration of brandy.

In India it is believed that no bite of the Hydrophis is known to have recovered.

Of bites by the Daboia (Crotalidæ), nine cases occur:

1	case of	7	hours.		1 case of	36	hours.
2	"	24	"	each.	1 "	2	days.
1	"	27	"		1 "	9	"
1	"	28	"		1 "	recovered.	

This gives 6 cases ranging from 7 to 36 hours, and one case of 2 and another of 9 days.

Although Dr. Fayrer classifies the poisons in the following order of deadliness: 1st, Cobra; 2d, Daboia; 3d, Hamadryad; 4th, Bung. cœrul. (Krait); 5th, Bungarus fasciatus; 6th, Echis; 7th, Crotalidæ; 8th, Callophis; yet, judging by the length of time in which death ensued, they stand thus, viz.: No. 1, Cobra; No. 2, Krait; No. 3, Hydrophis; No. 4, Daboia; and the cases marked unknown, afford no clue for classification. They vary from 12 hours, 18 hours, and 31 hours, to 42 hours; and from 2 days to 6 and 9 days. In 1 case the bite produced "no result;" and in 5 cases recovery ensued in respectively 24 hours, 2 days, 10 days, 13 days, and 22 days.

EXPERIMENTS WITH SNAKE-POISON.

The following experiments were made with snake-poison on animals; some trials of different antidotes are also given.

Experiment No. 1. A full-grown Pariah dog was bitten by a Cobra; died in 26 minutes, although previous to the bite he had been made to eat 5 leaves of the Aristolochia Indica.

No. 2. A fowl. Cobra poison injected into the thigh; died

in 3 hours, although it had eaten some of the leaves of the Aristolochia Indica. The blood of No 1 showed no changes under the microscope; but that of No. 2 contained several large granular bodies in distinct cells.

No. 3. A full-grown Ptyas mucosus (Rat snake), five feet in length, was bitten by a Cobra; no apparent result ensued.

No. 4. A full-grown Cobra was made to bite another Cobra; no apparent result.

No. 5. A full-grown Pariah dog was bitten by a Cobra; dead in 30 minutes.

No. 6. A pup, 3 months old, had a small quantity of Cobra poison injected into the thigh at 9.07 A.M.; sleep and drowsiness supervened, but at 2 P.M. he appeared quite well again.

No. 7. A fowl had a portion of the poison used in No. 6 injected into its body; dead in 1 hour.

No. 8. A pigeon had some of the same poison injected into the thigh; dead in 23 minutes. No alteration in the blood could be detected under the microscope.

No. 9. The pup of Experiment No. 6 had a larger quantity of the same poison injected into the other thigh at 9.02 A.M. At 11 A.M. drowsy; 12.30 vomits a dark-colored fluid; at 12.55 P.M. falls on his side and gasps for breath. Gave him 10 minims sulphuric ether; at 1.33 another dose of ether; at 1.45 pulse 175; 2 P.M. dose of ether; 2.36 pulse 26. Convulsions at 3 P.M. or in 6 hours, dead.

No change could be detected in the blood under the microscope with a $\frac{1}{25}$ and $\frac{1}{50}$ inch object-glass.

Action of Cobra Poison on Cold-blooded Animals.

Experiment No. 10. A full-grown (8 feet long) Ptyas mucosus (Dhamin), bitten in 3 places by a Cobra, two-thirds grown. No change could be detected at the end of three days, at which time the wounds produced by the bite had healed up; the snake died in 7 days (Dr. Fayrer adds) without any apparent cause! (?)

No. 11. A lizard, Varanus flavescens or Gohsamp, was bitten in two places by a Cobra 6 feet in length; dead in 3 hours 22 minutes.

No. 12. The Cobra used in Experiment No. 9 was bitten by a darker-colored Cobra; no apparent result.

No. 13. A Ptymuc (6 feet long) was bitten by the Cobra of Experiment No. 11; died in 46 to 48 hours.

No. 14. A large bullfrog (Rana tig.), was bitten twice by a Cobra; no apparent result.

In Experiments Nos. 10, 11, 13, no change in the blood could be detected under the microscope.

Action of the Poisons of the Cobra and Bungarus Fasciatus on Warm- and Cold-blooded Animals; also the Influence of Carbolic Acid on Snakes.

No. 15. A fish (Ophio. mar.), 14 inches long, was bitten near the tail by a Cobra; died in 50 minutes.

No. 16. A dog was bitten by a Bungarus fasciatus, full-grown, in two places on the thigh; died in about 74 hours.

No. 17. A young mongoose (Herpestes malacceissis) was bitten three times by a full-grown Cobra; died in 26 minutes; convulsions supervened in three minutes after the bite.

No. 18. A Ptyas mucosus, bitten by a large Cobra; no effect.

No. 19. A Ptyas mucosus, bitten by a large Cobra; no effect.

No. 20. A cat bitten by the Cobra of Experiment No. 17; dead in 1 hour 30 minutes.

No. 21. A dog, bitten by a Bungarus fasciatus, 6 feet long;

died in 2½ days, although on the day following the bite he seemed perfectly well and ate well.

No. 22. A cat, bitten by a Bungarus fasciatus, in the thigh; died the following day.

No. 23. A large mongoose, bitten by a large Cobra; no apparent effect.

No. 24. A Cobra, bitten by a large Bungarus fasciatus; no effect.

No. 25. A Cobra, bitten by another Cobra; no apparent effect.

No. 26. A mongoose, put into a cage with two Cobras; bit them and was bitten in turn; but when taken out on the next day one of the Cobras bit him on the thigh, when he died shortly thereafter.

No. 27. A large Cobra had a few drops of carbolic acid put into its mouth; a spasmodic action in the whole body ensued; dead in 20 minutes.

No. 28. A few drops of carbolic acid were poured on the floor of a cage containing a large Bungarus fasciatus; a drop fell on its head; it became convulsed, remained motionless for 5 or 6 minutes, and in 10 minutes death ensued.

Action of the Cobra Poison on Animals, and the Influence of Carbolic Acid on the Cobra and Frog.

No. 29. A Cobra was made to bite another Cobra in the head and mouth; no apparent result was produced.

No. 30. A Cobra was made to bite a large bullfrog; died in 1 hour 15 minutes.

No. 31. A large Ptyas mucosus was bitten by a large Cobra; no effect.

No. 32. A large Varranus flavescens, bitten by a Cobra in the mouth and thigh; died in less than 48 hours.

No. 33. A fowl had Cobra poison injected into its thigh; died in 4 minutes.

No. 34. A large bull had 10 drops of Cobra poison injected into his body; died in 28 minutes.

No. 35. A large Cobra was administered one drop of carbolic acid; convulsions in two minutes; at 4.34 apparently helpless; recovered in 24 hours.

No. 36. A small Cobra was administered same dose; died in 5 minutes.

No. 37. A large frog (Rana tig.) had 2 drops of carbolic acid poured on his head; died in 25 minutes.

No. 38. A large Ptyas mucosus had 10 drops of Cobra poison injected under the mucous membrane of the mouth; no apparent effect.

No. 39. A Paddy-bird (Ardea leucoptera) was bitten by a Bungarus fasciatus in the thigh; died in 1 hour 30 minutes; the bird was given to a wild cat, which ate it, without any apparent ill effect.

No. 40. Another Paddy-bird had poison of the Cobra injected in the wing; died in 20 minutes; the bird was given to a dog, who ate it, without any apparent result.

No. 41. A full-grown Pariah dog was bitten by a Daboia; died in 1 hour 13 minutes; this dog died without convulsions and without apparent pain.

No. 42. Another dog bitten by a large Daboia died in 1 hour 15 minutes.

No. 43. A kite was bitten by the same snake in the wing; appeared affected by the poison for an hour, but this wore off, and the next day it was well.

No. 44. A kite was bitten by a Daboia in the thigh; died in 19 minutes; no microscopical change could be detected in the blood.

No. 45. A Daboia, full grown, was bitten by a Cobra; no effect.

No. 46. A large dog was bitten by a Daboia in the hind leg (same snake of Experiment No. 45); died in 3 hours; a profuse mucous discharge from the stomach and of blood and mucus from the bowels after the bite. Three doses of a supposed antidote given at intervals of 7 to 10 minutes.

No. 47. A young pig bitten in the leg by a Cobra; died in 13 minutes.

No. 48. A Grass Snake bitten by a Cobra (No. 47); dead in 13 minutes.

No. 49. Two harmless Tree Snakes bitten by the same Cobra; died in 7 minutes and 9 minutes.

No. 50. A Dhamin bitten by the same Cobra; died in 6 hours 45 minutes.

On the influence of the Poisons of the Cobra and Daboia, and Strychnia as an antidote.

No. 51. A Bungarus fasciatus bitten by a Cobra; died in 29 hours; other experiments of one Cobra's biting another gave no result.

No. 52. A harmless snake (Dendrophis) bitten twice by a Cobra; died in 41 minutes.

No. 53. A harmless snake bitten by Cobra No. 51; died in 36 minutes.

No. 54. A large dog bitten by a black Cobra in the thigh; immediate injection of a solution of half a grain of strychnia near the bitten place; tetanic spasms in 5 minutes; died in 6 minutes.

No. 55. A large Pariah dog bitten by a black Cobra in the thigh; immediate injection of strychnia solution in bitten part; died in 6 minutes.

No. 56. A large cat was bitten by a Daboia; injections of solution of strychnia, as in previous experiments; died in 7 minutes.

No. 57. A large Rat Snake was bitten by a Cobra; died in 2 hours 27 minutes.

No. 58. A large Cobra had a solution of strychnia injected into its body; died in 14 minutes.

No. 59. A large Cobra had 15 drops of Cobra poison injected into its body; died in 29 minutes.

No. 60. A large Cobra had 15 drops of his own poison injected into its body; no apparent effect.

No. 61. A large Cobra had 12 drops of partly its own and partly another's poison injected into its body; died in 5 hours 40 minutes.

No. 62. A large Cobra had 25 drops of poison from another Cobra injected into the body 8 inches from the head; no apparent effect.

No. 63. A fowl bitten by a Daboia in the thigh; died in 1 minute 30 seconds; death was preceded by violent convulsions.

No. 64. A fowl had half a drop of No. 63 injected into its thigh; dead in 2 minutes 10 seconds; death was preceded by paralysis and lethargy.

Several more experiments of injecting Cobra poison into Cobras produced no apparent effects.

No. 65. A Cobra had an injection of 10 drops of carbolic acid in the neck; dead in 5 minutes.

Experiments with the Poisons of the Cobra and Daboia on horses and other animals.

No. 66. A large dog, bitten by a Daboia in the hind leg, showed signs of the poison in about 5 minutes; disinclined to move in 43 minutes; died in 1 hour 6 minutes; not the slightest change could be detected in the blood under the microscope.

No. 67. A dog bitten by a Daboia; died in 8 hours 11 minutes. The blood in both these experiments was perfectly fluid 14 hours after death.

No. 68. A pig bitten twice by a Cobra; died in less than 3 hours; blood was firmly coagulated 1 hour after death.

No. 69. A small Ptyas mucosus bitten by a Cobra; died in 21 minutes.

No. 70. A mare, with paraplegia, bitten by a Cobra; died in 1 hour 20 minutes.

No. 71. A horse, 27 years old, paraplegic; bitten by a Daboia; died in less than 12 hours.

No. 72. A bay gelding, paraplegic; bitten by a Daboia near the right jugular vein; died in 11 hours 45 minutes.

No. 73. A mare, 27 years old, paraplegic; bitten by a Cobra; died in 1 hour 20 minutes; no possible change could be detected in the blood under the microscope.

Other trials of Cobra-bites on Cobras and Daboias were made without effect.

No. 74. A fowl was bitten by a Cobra; died in 1 minute 40 seconds.

No. 75. Thirty drops of blood from fowl No. 74 were injected into the thigh of another fowl; death ensued in 3 hours 50 minutes.

No. 76. Another fowl had the blood of No. 75 injected into the thigh; no apparent effect.

No. 77. A fowl bitten by a Cobra; died in 50 seconds.

No. 78. A fowl had a syringeful of blood from No. 77 injected into the thigh; dead in 2 hours 46 minutes.

EXPERIMENTS WITH SNAKE-POISON. 169

Other experiments of Cobra-bites in fowls produced death in 8 minutes 15 seconds and 34 seconds.

Daboia-bites in fowls produced death in 45 seconds and 35 seconds.

No. 79. A cat had 5 drops of Cobra poison, diluted with 10 drops of water, injected into the leg; died in 1 hour 57 minutes 15 seconds.

No. 80. A fowl bitten by a Cobra was dead in 34 seconds.

No. 81. A fowl had two syringefuls of blood from fowl No. 80 injected into the thigh; died in 3 hours 33 minutes.

No. 82. A chicken had 25 drops of blood from No. 81 injected into its thigh; it died 9 days thereafter greatly emaciated.

No. 83. A fowl inoculated with Oph. elaps poison; died in about 45 minutes; the blood of this fowl injected into another fowl produced no apparent effect.

No. 84. A Cobra bit another Cobra which was afterwards bitten by a Daboia; it died in 3 hours 22 minutes.

No. 85. A muskrat had 4 drops of the blood of a fowl (which had been killed by a Cobra-bite) injected into its hind quarters; it ate a portion of the fowl from which the blood had been taken, and died in about 12 hours.

No. 86. A fish bitten by a Cobra; died in 20 minutes.

No. 87. A snail bitten by a Cobra; died in 32 minutes.

No. 88. Three chickens bitten by a Cobra; died respectively in 3 minutes 10 seconds, 5 minutes 30 seconds, and 4 minutes 30 seconds.

No. 89. A dog bitten by a Cobra; died in 1 hour 3 minutes.

No. 90. A dog bitten by a Daboia; died in 44 minutes.

No. 91. A fowl had one leg firmly bound by a ligature and was bitten below the latter by a Cobra; dead in 16 minutes,

No. 92. A dog bitten by a Bungarus fasciatus; died in 4 hours 28 minutes.

No. 93. A fowl bitten by the same snake; died in 26 minutes.

No. 94. Two fowls bitten by a Bungarus fasciatus; died respectively in 17 minutes, and 1 hour 55 minutes.

Several other experiments were made of Bungarus by Cobra, Daboia by Cobra, and Cobra by Cobra; without giving any apparent result.

No. 95. A full-grown dog bitten by a Cobra had one drachm of liquor ammonia diluted with water injected into the femoral vein; dead in 44 minutes 15 seconds.

No. 96. A large dog had 1 drachm of liquor ammonia, sp. gr. 959 B. P., injected into the femoral vein at 2.48 P.M.; at 2.49 howled loudly; cannot stand up; 2.50, convulsive twitchings in whole body; 2.57, lying quiet, no tremor; 3.13 P.M., sitting up apparently well.

No. 97. A large dog bitten by a Cobra, had 4 minims liquor ammonia injected into femoral vein immediately before the bite; in 30 minutes no apparent effect from the ammonia; had another injection; died in 48 minutes after the bite.

No. 98. A fowl had 20 minims liquor ammonia injected into femoral vein, and was then bitten by a Cobra; immediate convulsions and death.

No. 99. A fowl had liquor ammonia injected into the thigh; no apparent effect.

No. 100. A dog bitten by a Cobra, had 40 minims liquor ammonia injected into the left jugular vein; twenty minutes afterwards had 20 minims more injected; dead in 25 minutes.

Several further experiments of hypodermic injections of liquor ammonia, in cases of Cobra-bites, proved them of no efficacy whatever.

Injections of other Substances, Liquor Potassæ, Solution of Quinia, Cauterization, &c.

No. 101. A dog had an injection of 40 drops of Condy's liquor potassæ permanganatis* into the jugular vein, after which he was bitten by a Cobra; two more injections of the liquor were given, but death ensued in 37 minutes.

No. 102. A dog bitten by a Cobra in the thigh had 60 drops of a solution of quinia injected into the jugular vein; dead in 11 minutes.

No. 103. A pigeon had 15 drops of a mixture, composed of equal parts of Cobra poison and liquor ammonia, injected into the thigh; dead in 2 minutes.

No. 104. A dog had 10 drops of Cobra poison injected into the jugular vein; immediately thereafter an injection of 60 drops liquor ammonia into the same vein; dead in 1 minute 10 seconds.

No. 105. A dog had a leg ligatured, and was bitten by a Cobra below the ligature; the wound was immediately cauterized by red-hot steel points; dead in 1 hour 43 minutes.

No. 106. Another dog bitten and treated in the same way; died in 35 minutes.

No. 107. Five fowls bitten by Cobras; died respectively in 3 minutes, 10 minutes, 11 minutes, 17 minutes, and 22 minutes.

No. 108. Same Cobra bit a pigeon; died in 45 minutes; a fowl died in 1 hour 39 minutes; another fowl was bitten, fell, and was motionless for some time, but finally recovered.

In the preceding experiments, Nos. 107 and 108, the vemon was exhausted with the bite of the seventh animal.

No. 109. A fowl was bitten by a Daboia; carbolic acid was applied to the wound; dead in 1 minute.

---

* An injection recommended by Dr. Shortt of Madras.

No. 110. A fowl had a leg, bound by a ligature, bitten by a Cobra below it; carbolic acid was applied to the wound; no apparent effect from the poison at the expiration of 15 minutes; ligature was removed; died in 13 minutes thereafter, or in 28 minutes after the bite.

No. 111. A fowl had a ligature of a soaped cord tied around its leg, and drawn as tight as a strong man could draw it; the part below the ligature was bitten by a Cobra; fowl died in 44 minutes.

No. 112. A dog bitten by an Echis carinata; died in $9\frac{1}{2}$ hours.

No. 113. A fowl bitten by the same snake; died in 4 minutes.

No. 114. A fowl bitten by the same snake; died in 4 hours 7 minutes.

No. 115. A pup, 3 months old, bitten by an Ophiophagus 7 feet long; died in 1 hour 12 minutes.

No. 116. An old bull was bitten by a Cobra; the wound was rubbed with a fowl's liver, and a Tanjore pill administered; he recovered.

No. 117. A fowl bitten by the Cobra of Experiment No. 116; died in 10 minutes.

No. 118. A goat bitten by a Cobra had a fowl's liver rubbed on the bite and a Tanjore pill given; it recovered.

No. 119. A fowl bitten by the same Cobra had a Tanjore pill rubbed on the wound, and several other pills administered; dead in 6 minutes 40 seconds.

No. 120. A fowl bitten by a Hydrophis (salt-water snake); died in 14 minutes.

No. 121. Another fowl bitten by same snake; died in 17 minutes.

No. 122. Another fowl bitten by a Hydrophis; died in 9 minutes.

No. 123. A dog bitten by a Hydrophis; died in 1 hour.

No. 124. A fowl had the blood of No. 123 injected into its body; died in 12 hours.

Different kinds of poisons, namely, Cobra, Daboia, Ophiophagus, and Echis, applied to the conjunctiva of different animals, produced intense local inflammation, but in no case death.

The composition of the Tanjore pill* is as follows, viz.: Equal quantities of Arsenic (white).; Nervalam almonds (a narcotic); Root of Velli-navi (a purgative); Pepper; Root of Neri-visham (a purgative); Mercury. Mix and rub the Mercury with the sap of the Wild Cotton (Asclepias gigantea, *Linn.*), until the globules disappear entirely.

The dogs used in the preceding experiments were all full-grown Pariahs, a large bony animal common in India.

---

* Buffon's Natural History, vol. lxxxiii, p. 87.

Summary of the Preceding Experiments with Poisons.

Animals used.	Kind of Poison.	Caused death in	Animals used.	Kind of Poison.	Caused death in
Dog,	Cobra,	26 minutes.	Bull,	Cobra inj'd,	28 minutes.
Fowl,	" (Arist. ind.),	3 "	Cobra,	Carbolic acid adm.,	rec'd in 24 h.
Pty. muc.,	"	No result.	"	" "	5 minutes.
Cobra,	"	"	Frog,	" ext'l	25 "
Dog,	"	30 minutes.	Pty. m.,	Cobra inj'd,	No effect.
Pup,	"	Recovered.	Paddy-bird,	Bung. f.,	1 hr. 30 min.
Fowl,	"	1 hour.	"	Cobra inj'd,	20 minutes.
Pigeon,	"	23 minutes,	Dog,	Daboia,	1 hr. 13 min.
Pup,	?	6 hours.	"	"	1 hr. 15 min.
Pty. m.,	" ⅔ grown,	7 days.	Kite,	"	Recovered.
Lizard,	"	3 hrs 22 min.	"	"	19 minutes.
Cobra,	"	No result.	Daboia,	Cobra,	No effect.
Pty m.,	"	48 hours.	Dog,	Daboia,	3 hours.
Bullfrog,	"	No result.	Pig,	Cobra,	13 minutes.
Fish,	"	50 minutes.	Grass snake,	"	13 "
Dog,	Bung. f.,	74 hours.	Tree snake,	"	7 "
Mongoose,	Cobra,	26 minutes.	"	"	9 "
Pty. m.,	"	No effect.	Dhamin,	"	6 hrs.45 min.
"	"	"	Bung. f.,	"	29 hours.
Cat,	"	1 hr. 30 min.	Harmless sn	"	41 minutes.
Dog,	Bung. f.,	2 dys. 12 hrs.	"	"	36 "
Cat,	"	24 hours.	Dog,	"	6 "
Mongoose,	Cobra,	No effect.	P. dog,	"	6 "
Cobra,	Bung. f.,	"	Cat,	Daboia,	7 "
"	Cobra,	"	Rat snake,	Cobra,	2 hrs.27 min.
Mongoose,	"	Short time.	Cobra,	Strychnia inj'd,	14 minutes.
Cobra,	Carbolic acid,	20 minutes.	"	Cobra inj'd,	29 "
Bung. f.,	"	10 "	"	Cob. p. 15 dr. inj'd,	No effect.
Cobra,	Cobra,	No result.	"	" 12 dr. "	5 hrs.40 min.
Bullfrog,	"	1 hr. 15 min.	"	" inj'd,	No effect.
Pty. m.,	"	No effect.	Fowl,	Daboia,	1 min.30 sec
Lizard,	"	48 hours.	"	½dr. precg. one inj'd	2 min. 10 sec
Fowl,	" inj'd,	4 minutes.	Cobra,	Carbolic acid,	5 minutes.

SUMMARY OF EXPERIMENTS.

Summary of the Preceding Experiments—Continued.

Animals used.	Kind of Poison.	Caused death in	Animals used.	Kind of Poison.	Caused death in
Dog,	Daboia,	1 hr. 6 min.	Dog,	Amm. inj'd,	Recovered.
"	"	8 hr. 11 min.	"	Cobra,	48 minutes.
Pig,	Cobra,	3 hours.	Fowl,	{ Cobra, with inj'ts of liq. amm.,	1 minute.
Pty. m.,	"	21 minutes.	"	Liq. am. inj'd,	No effect.
Mare,	"	1 hr. 20 min.	Dog,	Cobra,	25 minutes.
Horse,	Daboia,	12 hours.	"	"	37 "
"	"	11 hr. 45 min	"	"	11 "
Mare,	Cobra,	1 hr. 20 min.	Pigeon,	Mix.cob. and liq. am	2 "
1. Fowl,	"	1 min. 40 sec	Dog,	Cobra inj'd,	1 min. 10 sec
2. "	30drs.of prec. one inj	3 hr. 50 min.	" ligature	" wound cautr'zd	1 hr. 43 min.
3. "	Blood of No. 2 inj'd,	No effect.	" "	" "	35 minutes.
"	Cobra,	30 seconds.	1. Fowl,	Cobra,	3 "
"	Blood of prec'g inj'd,	2 hrs. 46 min	2. "	Same cobra as No. 1,	10 "
"	Cobra,	8 min. 15 sec.	3. "	" "	11 "
Fowls,	"	34 seconds.	4. "	" "	17 "
"	Daboia,	45 "	5. "	" "	22 "
"	"	35 "	6. "	" "	45 "
Cat,	Cobra inj'd,	1hr. 57m.15s.	7. "	" "	1 hr. 39 min.
1. Fowl,	Cobra,	34 seconds.	8. "	" "	Recovered.
2. "	Blood prev. inj'd,	3 hrs. 33 min	"	Daboia,	1 minute.
3. Chicken.	" of No. 2 inj'd,	0 days.	"	liga. Cobr.,lig.rem. in 15m	28 minutes.
Fowl,	Ophioph. inj'd,	45 minutes.	" "	Cobra,	44 "
Cobra,	Cobra,	3 hrs. 22 min	Dog,	Echis c.,	9 hrs. 30 min
Muskrat,	{ Blood of a fowl killed by Cobr. inj.	12 hours.	Fowl,	"	4 minutes.
Fish,	Cobra,	20 minutes.	"	"	4 hrs. 7 min.
Snail,	"	32 "	Pup, 3 mos.,	Ophioph., 7 ft. long,	1 hr. 12 min.
Chicken,	"	3 min. 10 sec	Bull,	{ Cobra,fowl's liver and Tanjore pills.	Recovered.
"	"	5 min. 30 sec			
"	"	4 min. 30 sec	Fowl,	Cobra, fowl's liver,	10 minutes.
Dog,	"	1 hr. 3 min.	Goat,	" Tanj. pill,	Recovered.
"	Daboia,	44 minutes.	Fowl.	{ Cobra, Tanj. pill internal entirely,	6 min. 40 sec.
Fowl,	Cobra,	16 "	"	Hydroph.,	14 minutes.
Dog,	Bung. f.,	4 hrs. 28 min	"	"	17 "
Fowl,	"	26 minutes.	"	"	9 "
"	"	17 "	"	"	1 hour.
"	"	1 hr. 55 min.	Dog,		
Dog,	Cobra,	44 min.15sec.	Fowl,	Blood precg. inj'd,	12 hours.

Making in all 124 cases, which may be classified as follows, viz.: Cobras bitten by Cobras, 5 cases; 4, no effect; one case died in 3 hours 22 minutes. Cobras injected with Cobra poison, 4 cases; one died in 29 minutes, another in 5 hours 40 minutes; 2 cases recovered. Dogs bitten by Cobras, 11 cases, varying from 6 minutes to 1 hour 3 minutes; the greater part of the cases, however, average from 25 minutes to 45 minutes. Of dogs injected with Cobra poison, one died in 1 minute 10 seconds. Dogs that had the ligature applied and cauterized, one died in 1 hour 43 minutes, one in 35 minutes. Fowls died in from 34 seconds to 3 hours.

In three separate experiments fowls died from injected blood from fowls killed by Cobras; and in a third case a chicken died in 9 days from injected blood of a fowl that had died from injected blood of a fowl previously killed by a Cobra-bite.

This last case is one of the most interesting in the whole series of experiments, as it seems to accord with Prof. Halford's theory of the poison containing a germinal matter, and also of Dr. S. Weir Mitchell's theory of its nature as a ferment.*

Injections of liquor ammonia in fowls were not attended with success in any case. In cases of ligatures applied to the limb *before the bite* death occurred in 28 minutes, and 44 minutes, which shows that despite the compression of the arteries and veins in the limb, a ligature cannot be put on so tightly but that the poison will pass with great rapidity into the general circulation, and produce death.

Carbolic acid was given to a Cobra; it recovered in one case, but in all the remaining cases death ensued in 5 minutes, 20 minutes, and 25 minutes. The bull and the goat bitten by Cobras, which were treated by fowl's liver rubbed on the wound and Tanjore pills administered, both recovered, although a fowl treated in the same way died in 6 hours 40 minutes.

---

* See page 132.

MICROSCOPICAL APPEARANCE OF COBRA-POISON. 177

In experiments made by Dr. Russell with the Daboia poison it killed a dog in 26 minutes, and a chicken in 36 seconds. The *Cobra poison injected into the jugular vein* is not known to have produced death in less than 34 seconds.

Fig. 2 is from Fayrer's Thanatophidia, and is a sfigma taken by Dr. D. Douglas Cunningham, of Bengal. Under

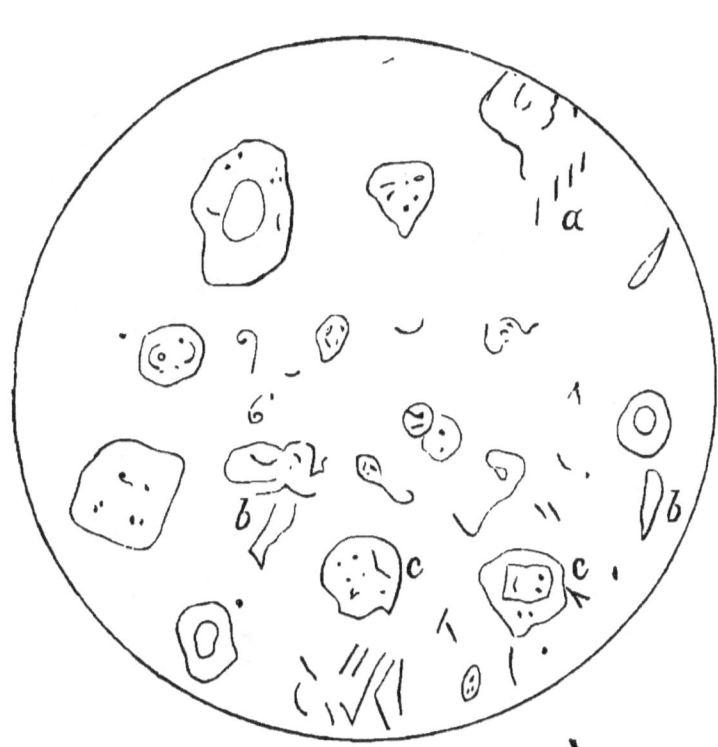

Appearance of the Cobra-poison under the microscope.

what magnifying power this appearance is exhibited is not stated.

The poison-globules have the general appearance of spotted disks and dots, lines and curves; a most peculiar obtuse angle, formed by the junction of two straight lines, and other groups of lines converging from common centres, suggest crystals, as shown in Fig. 13.

MICROSCOPICAL SFIGMAS, SHOWING THE APPEARANCE OF THE POISONS IN THE BLOOD OF A DOG AND OF A FOWL.

As taken by Dr. Douglas Cunningham, Bengal.

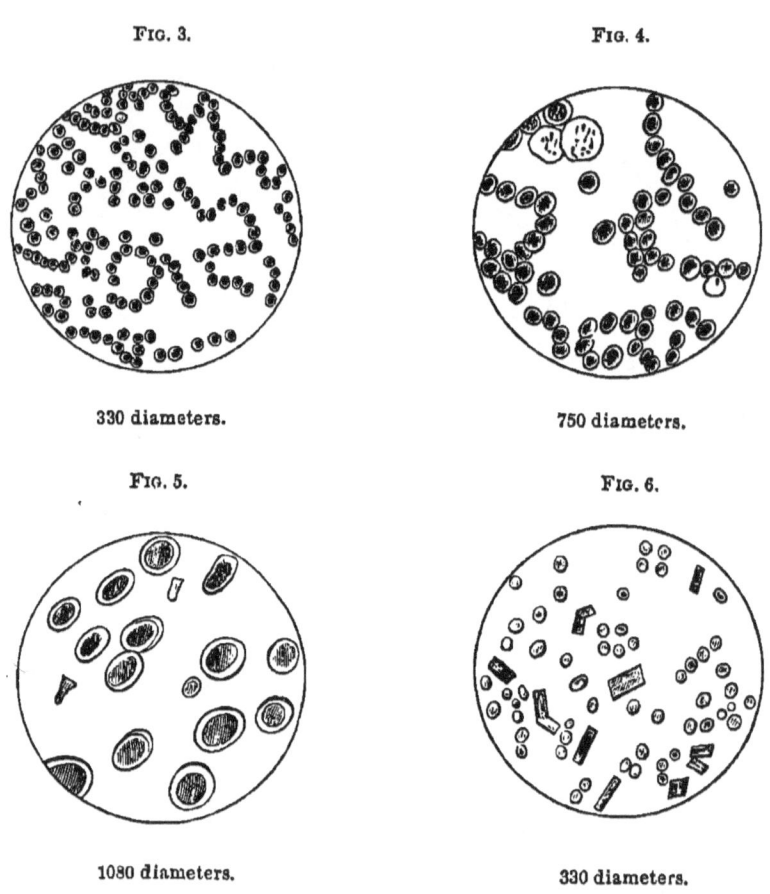

Fig. 3. 330 diameters.

Fig. 4. 750 diameters.

Fig. 5. 1080 diameters.

Fig. 6. 330 diameters.

Figs. 3, 4, and 5 show the microscopical sfigmas of *Cobra-poison* in the blood of a dog.

MICROSCOPICAL SFIGMAS OF COBRA-POISON. 179

Figs. 6, 7, and 8 show the microscopical sfigmas of the *Vipera Daboia Russellii* poison in the blood of a dog.

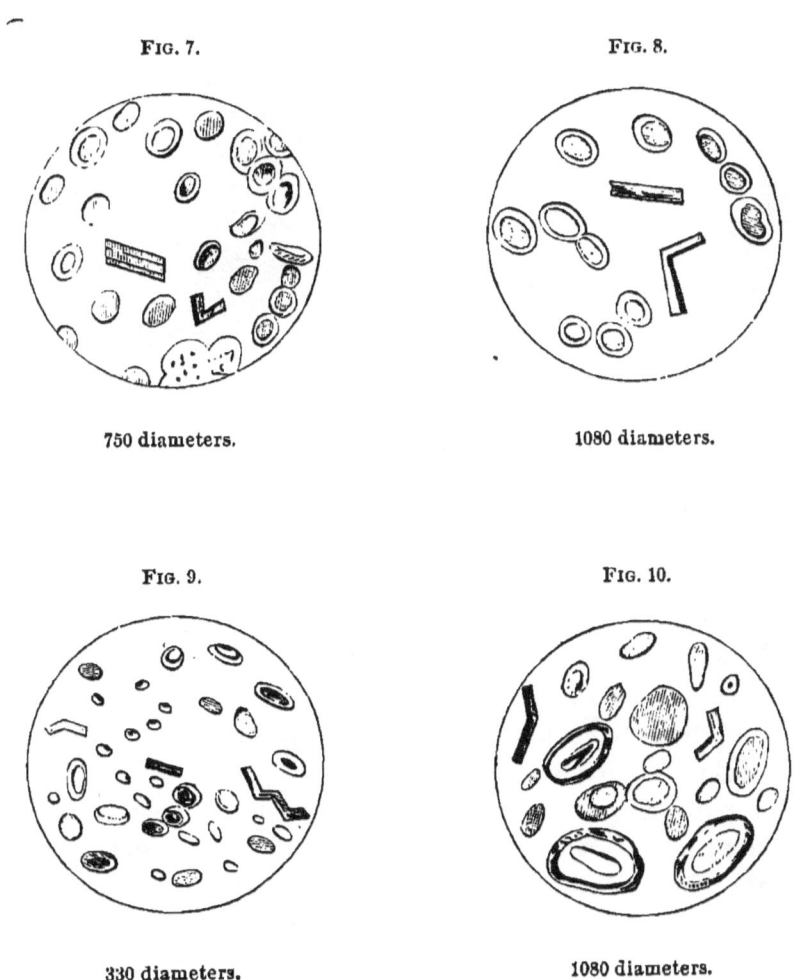

FIG. 7.

750 diameters.

FIG. 8.

1080 diameters.

FIG. 9.

330 diameters.

FIG. 10.

1080 diameters.

Figs. 9 and 10 show the microscopical sfigmas of the *Cobra-poison* in the blood of a fowl.

Figs. 11 and 12 show the microscopical sfigmas of the *Daboia poison* in the blood of a fowl.

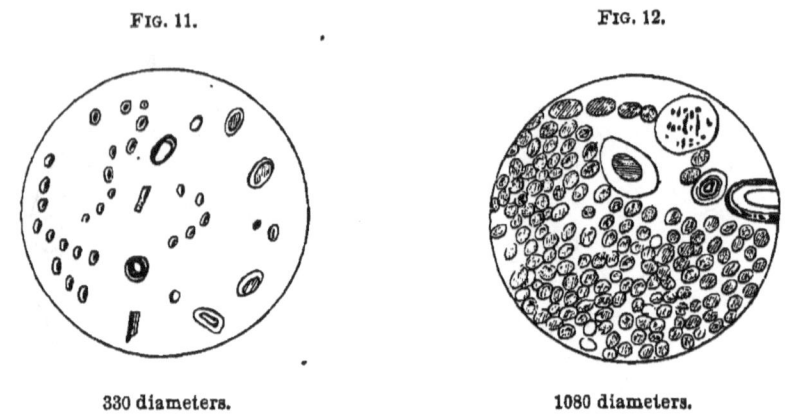

FIG. 11.   FIG. 12.

330 diameters.   1080 diameters.

Study of the Sfigmas.

Fig. 3 does not show a single poison-globule or corpuscle in the field. Fig. 4 exhibits two disk-like corpuscles, with an adjacent blood-globule which has *a double-lined* periphery, and several others of a similar appearance in other parts of the field; a greater proportion of the blood-corpuscles remain unaffected. In Fig. 5, under a power of 1080, only three or four of the blood-globules are not changed in appearance, and the oblique figures are wanting. Figs. 5, 6, 7, and 8 all exhibit parallelograms and the oblique figures under the three different powers used, but only Fig. 7 shows a poison-globule in the field; each one, however, has corpuscles evidently changed by the poison, as indicated by large numbers of them with the double-lined periphery. In Figs. 9 and 10 the poison-corpuscles are much more abundant, and the appearance indicates a much greater decomposition of the blood than in any of the other sfigmas. The poison-corpuscle in Fig. 12, although of the same kind as in Fig. 7, yet has a regular periphery like that of the Cobra-poison in Fig. 4. The periphery of that seen in Fig. 7 is irregular, but still so similar to certain disks in Fig. 2 in appearance, that the only *apparent* difference in the microscopical appearance of the two kinds of poison is the close grouping of small corpuscles seen in Fig. 12, which is wanting in all the others, and only seen under a power of 1080.

The following, Figs. 13 and 14, are from Dr. S. Weir Mitchell's "Researches upon the Venom of the Rattlesnake."

Fig. 13. These crystals were obtained by diluting the venom and allowing the mixture to dry slowly, sheltered by a cover-

FIG. 13.

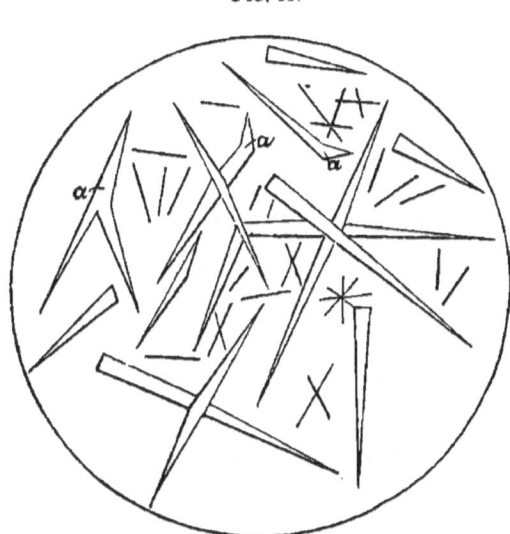

Crystals deposited from the diluted venom of the *C. confluentus*, by Prof. Hammond.

glass. The crystals thus formed resemble those of ammoniaco-magnesian phosphate, which affect the feathery form of crystallization.

The white deposit was composed chiefly of amorphous,

granular matter, with a few pavement epithelial cells, compound granular bodies of an oleaginous character, and finally of the peculiar masses known and described as colloid bodies,* and in appearance so much resembling starch granules as to have induced me to neglect them at first, supposing them to be really that substance accidentally present. These corpuscular bodies were marked with delicate radiating lines. Iodine stained them a yellowish-brown. They were doubtless due to some concrete modification of albuminous material.

Occasionally, when the snake had been seriously maltreated, the venom contained more or less blood. This sfigma serves to elucidate certain appearances in Fig. 2: a, a† are masses of crystals which, under a higher power, would present a similar appearance to the field in Fig. 13; b, b† are similar groups, somewhat distorted; while c, c† are poison-corpuscles or disks, upon which are seen the peculiar oblique-angled crystals marked a, a, a, in Fig. 13.

Fig. 14 shows the ultimate effect of the venom upon muscular fibre, made apparent by disturbing it mechanically. Dr. Mitchell continues: "The final influence of the venom upon muscular structure was extremely curious. In every instance it softened it in proportion to the length of time during which it remained in contact with it, so that after even a few hours in warm-blooded animals, and after a rather longer time in the frog, the wounded muscle became almost diffluent, and assumed a dark color and somewhat jelly-like appearance. The structure remained entire until it was pressed upon or stretched, when it lost all regularity, and offered the appear-

---

* Wedl, Pathological Histology, Trans. of the Sydenham Society, pp. 38, 264, 271.

† Mitchell's Researches, &c.

ance under the microscope of a minutely granular mass, dotted with larger granules, as shown in Fig. 14."

FIG. 14.

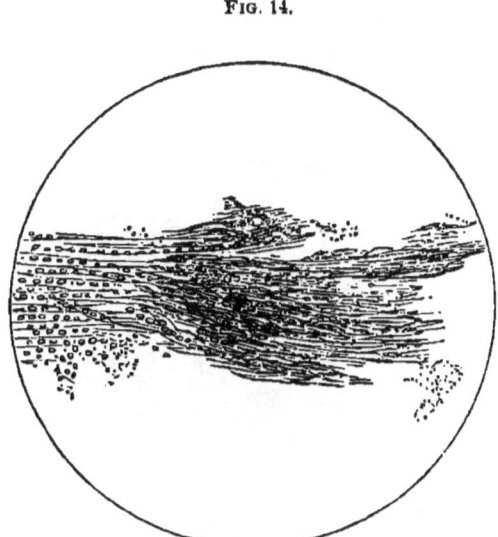

Appearance of muscular fibre after contact with venom. By S. Weir Mitchell, M.D.

Structure of the Venom-Gland.

The sfigma shown in Fig. 15 gives the microscopical appearance of the structure of the venom-gland.

The cœca lie in the centre, and the ducts are disposed on either side parallel to each other.

Outside of the cellular layer, the poison-duct is made up principally of white fibrous tissue, with a small proportion of very fine fibres of yellow elastic tissue. The walls of the duct are provided throughout with an abundant supply of blood-vessels. Its communication with the fang and the mode of injection of the poison have been described on another page, and any one desirous of knowing all the details of a minute description of these parts is referred to Rymer Jones, article "Reptilia," *Cyclopædia of Anatomy and Physiology;* J. L. Soubeiran, "De la Vipere," Paris, 1855; and Owen, on the "Skeleton and Teeth," Philadelphia, 1854. A consultation of these works, in connection with Dr. S. Weir Mitchell's

Fig. 15.*

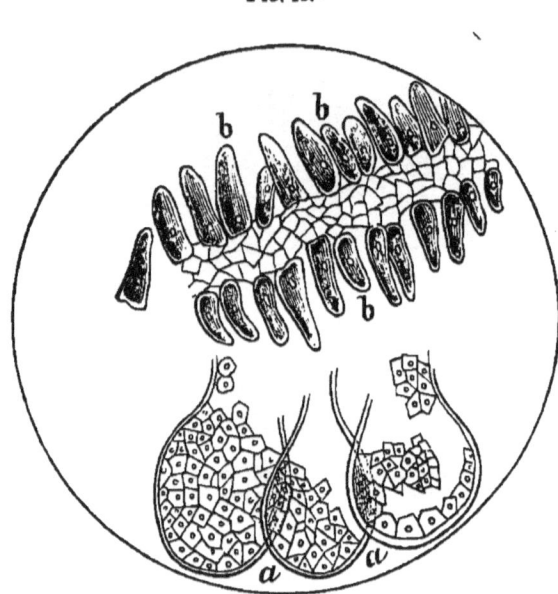

a, a. Secernent cæca. *b, b, b.* Small ducts.

"Researches," will furnish an exhaustive study of the subject in reference to Ophidians provided with fangs.

I intend to publish, at a future day, a like study of the same apparatus in poisonous serpents not provided with fangs; such, for example, as the *Vipera elaps corallinus.*

It must always be borne in mind, however, in this connection, that there is a wide difference in many respects between

---

* Researches, &c., by Dr. S. Weir Mitchell, p. 13.

poisonous reptiles in temperate zones or cold climates and the same species in torrid zones. In the tropics, on the Andes Mountains, for example, this fact is so plainly evident that, by referring to the tabular list of snakes in Colombia, where the same species occur first in some States, they occupy a second, third, fourth, or fifth rank in other places or districts. The cause of this is that in different localities the same snake occurs in different isothermal zones. A Taya (*Vipera pseudechis major*), for instance, found at a height of 200 metres above the sea-level, is *much more venomous and invariably holds a greater quantity of poison in deposit*, than one found on the Plains of Bogota, or at an elevation of 2300 metres. The one would have its corresponding parts so arranged and disposed as to hold a comparatively large quantity of venom in a poison-bladder; while the other, requiring and using a much smaller quantity, would have the duct from the poison-gland to the fang slightly enlarged near its attachment to the base of the latter, and consequently the poison-vesicle would be wanting.

ANTIDOTES.

The following list of substances used by the Curers in South America is copied from manuscript, and, together with the "methods of cure," these have been handed down from one to another, in all probability for more than two hundred years. Their use is of course purely empirical, but based upon long years of study and experiment by men who made these observations a specialty; many of them are ignorant, unable to read or write, but occasionally one of their number possessed these acquirements, and he pens down such additions or amendments as some venerable Curer may dictate to him.

They keep their secrets among the fraternity, exercising a sort of freemasonry, to which no one is admitted unless he give proofs of considerable knowledge of what they call "brujería" or incantations.

My knowledge of animal magnetism and physics enabled me to acquire quite a reputation among the common people as a "brújo," which I profited by to gain the respect of the Curers; and after trying experiments with snakes in their presence several times, I was solicited "to submit to a test," to which I consented.

They have knowledge of the preparation of a powder whose principal ingredient is made from *the spawn of a frog* (Bufo sahytiensis), minute doses of which will produce death slowly but surely; and as yet no autopsy or chemical analysis has been able to detect the substance or determine the cause of death in such cases.* They know an antidote for this substance, which is prepared by mixing other ingredients, of little or no medicinal virtue, with the gall of the frog, or that of the *Vipera acuaticus carinata*, or sometimes both. I had tested the powder and the antidote on dogs, and with success, and always carried the latter with me, with the expectation of having to use it on myself some time.

I had reason to suspect this substance might be given me. Nothing was said as to when or where the test would be made, but one day I was called to see a cure of a snake-bite performed by a celebrated Curer, and after the performance I was asked to his house, where dinner was served up to us. A glance of intelligence which passed between the man and his wife, who at the time placed two plates of soup before us, seemed to me to indicate my own dish as having been specially prepared. To call the old man's attention to a trifle in the corner behind him, with the utmost apparent indifference, enabled me to perform a sleight of hand trick in changing plates, and a moment after I was savoring my plate of soup quite sure that I had nothing to fear in partaking of the last drop of it. The Curer's eye sparkled and lighted up in quite an unusual way, only explained by the intense interest with which he was "studying symptoms" in myself.

Soup and dinner over he proposed a game at cards, and we were soon deeply interested in the charms of "retrúco," but after half an hour coffee was brought in, and this put an end to our "little game."

His cup of the beverage had hardly passed his lips ere he called for a lemon and some water, and inadvertently complained of a slight burning pain in the stomach. He went

---

* Is not this the substance of which so much has been said and told, which the Oriental doctors are said to prepare?

out of the room, and I overheard him ask the wife "if she was sure that I had the right plate of soup set before me?" "Of course she had;" so he returned and invited me to take a half hour in the hammock, which invitation I accepted, and while I appeared to sleep, the "study of symptoms" rather inclined to my side of the question. Severe colicky pains were soon present and caused so much alarm, that I was called to the side of the old man's couch to hear him say, "For the love of God, what is the matter with me?" I replied, "Nothing, of course, unless the plate of soup he intended for me had something in it, otherwise if anything had been put in it he ought to know what it was, for I had changed the plates while his back was turned, and he had eaten all the soup intended for me." I advised him "to take something" without delay, and left the house, sending back to him a brother Curer, and after two days he was out of danger. I was soon thereafter admitted to the "secrets of the brotherhood," and thus obtained their secrets, which I was only to divulge for a handsome remuneration. These are the "secret methods of cure" which are published in continuation.

SUBSTANCES USED AS ANTIDOTES BY THE CURERS.

No. 1. La Jenerala,—a climbing plant.
    2. La Capitana cascabel,—a climbing plant.
    3. La Capitana chifladora,—a climbing plant.
    4. La Cruceta de garfio,—leaves of a tree.
    5. La Cruceta palo de yuca.
    6. The fangs (pulverized) of the most poisonous snakes, together with the bones of the same snake's head, all well mixed together and added to the "Liga" (see page 190).
    7. Guaco de hoja peluda,—smooth-leaved guaco.
    8. Guaco morado liso,—smooth brown-leaved guaco.

ANTIDOTES. 189

No. 9. Guaco lisode hoja blanenzca,—white-leaved guaco.
10. Contra de la torcoral,—a plant.
11. Valdivia (la nuez),—a nut.
12. Sambita,—a plant.
13. Negrita,—a plant.
14. Plátano,—the plantain.
15. Cepa de guineo,—stalk of the banana.
16. Cedron (la nuez),—a nut (febrifuge).
17. Canelito,—bastard cinnamon.
18. Guanabanito,—a plant.
19. Annazuco,—a plant.
20. Carnestolenda,—a plant.
21. Escobilla menuda,—a plant.
22. Vara de sapo,—stalk of a plant.
23. Raiz del cojon-de-fraile,—a root.
24. Santa Maria,—a plant.
25. Contra-yerba,—a long grass.
26. Pepa del nispero,—kernel of the medlar.
27. La fabilla,—a stone of a fruit.
28. Colmillo de caiman,—alligator's tusks pulverized.
29. Cuerno de ciervo,—stag's horn.
30. Suelda con suelda,—a plant.
31. Ruda,—Rue.
32. Polvo de pezuña de ganado masho, tostada.
33. Mostaza,—mustard.
34. Azufre,—sulphur.
35. Arponcito ; alconcito,—Aristolochia Colombiana.
36. Solobasta,—Aristolochia Colombiana.
37. Estrella,—a plant.
38. Chupadora,—a kind of yuca.
39. Canela de sosa,—cinnamon.
40. Pepa de burro,—a fruit kernel.
41. Ojo de buei,—a fruit kernel.

Tinctures of these all mixed together form what is called "Liga," or "The League;" the most bitter antidotes are considered the most efficacious.

The plants named are found in certain localities in great abundance, and those considered most efficacious by Curers in certain places are always found in places infested by the most venomous snakes.

The same substances are called by different names by the different Curers, although scores of them have a manuscript list, of which the above is an exact copy, except that the botanical characteristics of the plants are written out in detail.

The Curers are generally men void of all principle; shrewd; cunning; ever ready to profit by any circumstance which can be turned to their advantage, and as they have to do with a people always inclined to superstitious belief; who attribute everything to "La Suerte" (luck, chance, fatality); as an inevitable result, these men are in a position to make themselves masters of any and every situation. They never forget an insult or injury given (except in a fellow-curer or "brujo,"), and will wait patiently for years with the hope that sooner or later they shall be called upon to cure a snake-bite in their enemy, and if death is not imminent they soon make it so by the administration or application of poisons they keep on hand, and their long-treasured insult is revenged by death.

They pass a great portion of their time in the woods and forests; find the hiding-places of the different snakes, and study their habits and peculiarities; thus it is that their knowledge is acquired from Nature herself, and not to be despised, though purely empirical.

I have succeeded in inducing many of them to use the

galls of serpents, prepared and administered as recommended on another page, and although they necessarily cling to many "hocus pocus" manipulations and performances, yet now they boast of never losing a case!

The first edition of the present work, published in 1870, in Bogotá, serves many of them as a manual, and has initiated a system of treatment in cases of snake-bites which is gradually spreading itself throughout the entire country.

Some of the tribes of the North American Indians use the galls of serpents (mixed with decoctions of herbs of little or no medicinal virtue), for curing bites; and the same may be said of the San Blas (Isthmus of Darien) tribes, and also of tribes in Venezuela, and in the valley of the Amazon.

A singular fact appears to be recognized in common by Curers and the "medicine men" among the Indians.

When performing a cure, they are always particular to have *an old woman* wait upon the patient, and explicitly forbid and prohibit any young woman or *a female whose menstrual flow is present* from approaching his bedside, *as this is known to cause an aggravation of the pains and augment the flow of blood*, if any hemorrhage exists! I can attest to having witnessed the truth of this fact in a great number of cases.

There is some psychological influence, some profound, unexplained relation, existing between the menstrual flow in woman and the serpent. I have seen in two cases of pregnant women, and in one case of a young woman whose menstrual flow was present, who stepped over a snake which was crossing the road, by mere chance, while I was following 'a few rods in the rear, came up to, and found the reptile winding and coiling itself up, as though in great pain, soon cease the violence of its movements, and appear to lose all power of locomotion, remaining semi-torpid for nearly an hour thereafter! This is a fact attested by hundreds of Curers, and by as many persons of undoubted veracity, to which I add my own evidence, without a shadow of doubt or hesitation.

Will Dr. Holmes, who has woven such a fascinatingly singular romance around psychological facts of a similar nature in his novel, entitled "Elsie Venner," please use his "deep-sea lines" of mental acumen, and tell us "the reason why" of this fact? A common element in the two beings is animal magnetism! I believe few physiologists deny its having some intimate and powerful influence upon the menstrual flow.

The snake, or its kind, is undoubtedly the most highly charged with magnetism, or animal electricity, of any other known reptile or animal, simply because its locomotive movements are those most peculiarly adapted to the development of this physical agent. Can it not be possible that this same agent is more closely allied to (is it not, possibly, one and the same thing as) *the vital force itself,* than is generally accepted as a creed in physiology? Does the larger quantity of magnetic fluid in the woman possess a *positive* character, and discharge itself upon the weaker and *negative* fluid of the reptile, causing a shock which stupefies it for a time? Does not a *like shock cause conception between the male and female?* Is this magnetic fluid the vehicle through which vitality is communicated to the germ which develops into a fœtus? Who will answer these queries?

Let us examine into some of the properties of the galls of serpents.

A chemical analysis of these would probably give a common element as a base, although the human bile is known to consist of cholic and choleic acids and cholesterin.

The poisons, as far as analyzed, give each its characteristic element for a base.

This appears to be further indicated in the fact that a preparation of the gall of a very venomous variety serves as a *perfect* specific for the poison of its own class, and also for all those varieties *less venomous;* although, in the latter case, recovery is not so prompt, and relief not so soon afforded, as when it is the gall of its own kind.

The gall of the most deadly kinds is very efficacious in all cases where the symptom it develops is known, and a corresponding one is noted in the patient, when its administration affords relief as if by magic. Thus, for example, the sting or bite of the Centipede causes excruciating, throbbing pains, accompanied by excessive œdema in the bitten limb; this finds precisely similar symptoms in the poison of the Vipera Acuaticus Carinata; the gall of which cures Centipede-bites in an hour! I have used the B. Lach. bufocephalus in many cases where "the kind" was doubtful, and with perfect success.

Drs. Buitrago and Lopez Z., of Barranquilla, at my suggestion, have used the B. Crotalus horridus for all kinds of wounds produced by poisonous reptiles; and both of them assure me that this has met with perfect success in every case, although the vicinity of Barranquilla is not abundant in snakes, of what are termed a "bad" kind.

A plant used by the Curers as an antidote and for inoculation, which I have classified as the Aristolochia Colombiana, merits more than a passing attention; its characteristics differ considerably from the Aristolochia Milhomens of Brazil.*

The tincture of this plant (root and leaves), is used by many of the Curers as a prophylactic for inoculation, and proves so specific a preventive against the poison as to enable them to perform many feats like the most famous snake-charmers of Oriental climes. The flower is singularly beautiful; the superior part is pear-shaped, pendant, terminating in a narrow

---

* See **Pathogenesie Bresilienne**, by Dr. Mure.

neck, which latter expands and forms the perianthium, gaping, ringent; the lower leaf (2) of which is prolonged to 25 centimetres in length, is funnel-shaped, terminating in a mere point, with both edges scolloped and contracted like a ruffle; the superior part is hermetically closed, and as the flower is being developed, it swells and puffs out, and when the pollen is fully ripened, the former bursts with force, making a noise like the lowing of a calf, and scatters the latter in every direction.

The most perfectly formed individuals of this species are found at from 200 to 300 metres above the level of the sea. They abound in the whole valley of the river Magdalena in cleared patches of land, and are most abundant in the plantain fields.

The Curers give it different names, the most common of which are Arponcito, Alconcito, Solobasta; the color of the flower is magenta brown articulations upon a reddish-white ground.

This is undoubtedly one of the most efficacious antidotes in common use. The root has a strong smell and taste of camphor. I prepared a homœopathic tincture of both the root and leaves, and with this inoculated several dogs, by making a slight incision with a lancet on the inner side of the fore leg, close to the shoulder-joint, introducing into the incision a small pellet of cotton saturated with the tinct. Arist. Col., one drop; some inoculations were made with an incision under each leg, others with only a single incision. The latter plan appears to be perfectly efficacious.

The animal falls into a semi-comatose state; in some cases a slight epistaxis supervenes, accompanied with bilious evacuations, dysury, slight œdema of the joints, particularly in the region of the incision; this state continues by intermissions for from four to ten days, after which the symptoms diminish gradually, and disappear entirely in from twelve to eighteen or twenty days.

During the period of inoculation *great care must be taken not to wet any part of the body*, and the Curers only allow drinks of boiled water, cooled and sweetened.

This period of prophylactic inception may be fixed at twenty days in a tropical climate, although in different climates its length must undoubtedly vary, and can only be determined by repeated experiments.

The dogs I inoculated in the town of Norosi would, after inoculation, tear any of the most venomous snakes in pieces, and if bitten, appeared to suffer little or no inconvenience from the bite, which healed of itself in three or four days; not one of them has, however, been killed by a snake-bite, when previously almost every dog bitten by a snake died.

I inoculated myself, by making incisions in the lower extremity of each deltoid muscle, with a preparation of the gall of the Lachesis. (B. Lach. one-tenth, or one drop of pure gall to ten drops of 95 per cent. alcohol.) The symptoms experienced were fugitive pains of different degrees of intensity in different parts of the head; slight epistaxis; adipsia; anorexia; dysury; diarrhœa,* which lasted four days; rheumatic pains in joints and limbs; cough, soreness of the throat, accompanied with painful deglutition; accesses of erotism; great debility of body and mind; vertigo; paleness of face and skin, and inclination to fainting fits at times.

These symptoms diminished in intensity after six or eight days by intermissions, and finally ceased one by one, till after twenty days all that remained was a strange taste of the leaves of a plant (Arist. Col.) upon the tongue, which was noticeable for more than a month afterwards.

---

* For fifteen days previously I had been much constipated.

Since being inoculated, I have never seen any snake make the slightest attempt to bite me, unless provoked, and I have observed the same fact in all the Curers, and every person inoculated; one is also freed, by inoculation, from the annoyance of flies, fleas, and sand-flies, which is by no means an inconsiderable relief in any tropical climate.

The duration of this preservative effect by inoculation is undoubtedly the same as that of the Virus vacuno.*

Another of the species Aristolochia is the Virginia snake-root[†] (Serpentaria or Aristolochia Virginia). It is not improbable that the Indians used it as an antidote, and that this fact originated its name.

The *Cedron Nut* (Simaba Cedron) is another substance frequently applied (in powder) to the bitten part, and administered internally in a decoction, but with little apparent effect in most cases, and of *none whatever* in many; the Cedron is the popular febrifuge.

The *Valdivia Nut* (produced by a vine like the poison ivy) is highly recommended by many Curers in the State of Antioquia, to be applied and used in the same way as the Cedron. My experiments with it do not warrant me in considering this antidote but slightly superior, as such, to the previous one.

Its action in antidoting the poison is slow in all cases, and ineffectual in very many, and it is by no means a specific.

The *Guaco* (*Mikania Guaco*) (leaves and root) is *said* to be used as an antidote by some of the Indian tribes in the interior of Colombia, and consequently enjoys a wide reputation throughout all classes in the country;[†] that it has cured some cases cannot be denied, but it is none the less true that several cases in which it has been administered have terminated fa-

---

* It is a singular fact, that no person inoculated is known to have died from any fever of a typhoid or putrid type.

[†] Jahr and Catellanus, Pharmacy, 1860.

tally. The Guaco does not seem to be more unfortunate as a specific than a thousand other remedies and patent medicines.

There are several varieties of this plant, known as the Morado (purple), Blanco (white), Cruceta (so called from the arrangement of its stems in the form of a cross), and the Real (royal). Its juice is very bitter, and it is often used by the country people as a febrifuge and vermifuge, with apparently good effects in some few cases.

Respectable authorities* advise the use of Spirits of Ammonia. I have never known of its application in a single case to have been followed by any apparent benefit. To test its virtues I made the following experiments: A female of the Mapana Saps (Lachesis) species, which was caught alive in Norosi, was allowed to bite a pup of 3 months old, which died in 5 minutes. Another pup of the same litter was presented to the reptile, and was bitten twice; Hartshorn was instantly applied to the wound, and a small quantity (10 drops) was administered in half a tumbler of water, but death ensued in 10 minutes. The day following another pup of the same litter was caused to be bitten, and after allowing *one minute* to elapse, that the poison might take full effect, one drop of Bungarus Lachesis, one-tenth, was administered in a small quantity of water, and one drop of the same antidote was put upon *one* of the wounds after opening it crosswise with a lancet; in an hour's time the pup was frolicking about as though nothing had happened it, and three months afterwards it was still alive and well.

The day following our last experiment this same snake bit a bitch that had had four or five litters of pups, and she died in less than an hour after receiving the bite.

---

* See page 143 for report of cases cured by Prof. Halford.

I leave every reader to make his own reflections upon the preceding experiments.

The celebrated South American savant, De Caldas, in his writings on "The New Kingdom of Grenada" (1807, 1808) says:

"A Noanama Indian, with whom I travelled through the upper country (Cundinamarca and the Cauca), showed me several plants used by his tribe for curing snake-bites. I observed that almost all of them belonged to the species Besleria."

Several other antidotes used by the Curers might be spoken of in this connection, but as another part treats especially of their different modes of treatment, those will be referred to in the different methods.

Dr. Vargas Reyes, of Bogota, has lately renewed the "Alcohol treatment," and calls it "specific," recommending it for general use. It can be safely considered a *good* treatment if adopted in time, or with little delay, but several cases in which it has been applied have terminated fatally, and I will cite a case which occurred many years ago in the northern part of the State of Ohio, which some of my readers will doubtless remember. It occurred about the year 1845, in Huron County.

A blacksmith, generally drunk and exceptionally sober, and therefore a splendid subject for the "Alcohol treatment," who had his shop at a cross-roads, went to his forge one morning to work, and found a large Rattlesnake coiled up upon his anvil; the morning "smile" which he had undoubtedly taken was quite enough to put our knight of the hammer and tongs in a fighting mood, and no helmeted knight ever took up a challenge to enter the lists more readily than did he the hissing challenge of his snakeship.

Seizing his hammer, which was doubtless the weapon he

best knew how to use, the fight began; how long it lasted, whether any of the rules of the prize ring were broken, never "got into the papers," but the son of the smith coming to the shop found the fallen knight of the hammer on the ground, and his snakeship wound about his neck, head erect, and smeared with the blood which was flowing from the wounds on the smith's face; the boy alarmed his mother, who entered the lists with a hickory broom-handle, and aided by an elder son succeeded in killing the snake. The father, who was bitten in many places about the head, neck, and breast, was removed to the house, and after the application of herb-tea and the whole list of household remedies, was restored to consciousness after three or four hours had elapsed; but for years afterwards he still suffered pains in the joints of a rheumatic nature, which made him almost helpless, and were followed by an epistaxis of three or four days' duration. These pains, &c., were felt *invariably at the period of the full moon*.

I met him in 1854 in Western North Carolina, and had the preceding statement from his own lips. A few doses of the B. *Crotalus horridus*, 1st decimal potency, 10 drops in 4 oz. aquæ destillatæ, a tablespoonful night and morning, cured him radically of all his sufferings.

This was an extreme case, but one can hardly imagine the "Alcohol treatment" applied under more advantageous circumstances *if the snake's poison had its deadly principle fully developed at the time of its injection.*

Dr. S. Weir Mitchell reduces the list of antidotes which still hold repute to: Ammonia, Olive oil, Arsenic (as the Tanjore pill), Bibron's antidote (Bromine), and alcoholic stimuli.

Although Ammonia has only proved worthy of note in the cases cited by Prof. Halford, yet the latter feels sanguine it must prove efficacious in many cases. In my own experience it has proven uniformly unsuccessful. Dr. Fayrer's experiments with it gave negative results, and Dr. S. Weir Mitchell was compelled to consider it of no value whatever as an antidote.

Injections of Bromine have been successfully made, and they also have been negative in results.

Injections of Iodine, although highly recommended under the patronage of the Smithsonian Institute, have given no better proofs of efficacy in counteracting the effects of the poison.

Dr. S. Weir Mitchell says: "Profound drunkenness is a condition of sedation and not of excitement, and yet the whole object of using Alcohol in snake-bites has been among rational men to stimulate and not to lull or depress the system. In fact it is well known that persons who were at the time dead drunk, or nearly so, have been bitten by Rattlesnakes, and have obtained thereby no immunity from the effects of the bite."

Dr. Brainerd, who is opposed to the use of stimulants in Crotalus-bite, thinks the evidence in its favor insufficient, and thus sums up his argument against its utility: "When mixed with alcohol the venom is rapidly fatal if inoculated."

Dr. S. Weir Mitchell concludes by recommending the inhalation of fumes of warm alcohol or ether.

An interesting fact in connection with the use of the Tanjore pill is that it is always recommended to be accompanied by *rubbing a fowl's liver on the bitten part!* Who shall say that the *bile* absorbed into the wound from the liver did not antidote, partially or wholly, the venom in the blood, and for which virtue is accredited to the arsenic of the "pill?"

The Indians in the Chocó (valley of the River Atrato) district use most frequently a plant called Flor de Gallo, Cocksflower; its botanical characteristics are:

Shrub one metre in height, stem erect, jointed, composed of a cylinder of greater diameter surrounded by four cylinders of smaller diameter in the shape of a Gothic column; fifteen centimetres long between the joints, from which spring the smaller lateral stems in pairs; alternate; terminating in a single leaf like that of the rose geranium; color of the leaves bluish dark green; hirsute on their superior surface, and wrinkled. On their lower surface bluish light green; stem and leaves of the same color; a single flower on the extremity of the central stem; its color bright velvet red, smooth; petals four; circumvolute; overlying each other; subcampanulate; stamens 0; pistils 1; 0.01 centimetre high, terminating in a diminutive acorn-shape, covered with pollen of a much brighter red than the petals; stem, branches, and lower surface of the leaves covered with exceedingly diminutive globules of a colorless, transparent, mucilaginous liquid, which rubbed between the hands exhales a pungent, resinous aroma that causes a momentary sensation of faintness, and seems to pass directly through the head from the nostrils to the cerebellum.

A tincture (alcoholic) made from the stem, leaves, and roots of this flower is used for curing; and if a snake is seen in the roof,* or suspected to be about the house, the natives place the bottle uncorked in the centre of the room where it was last seen, and abandon the house for a short time. The reptile comes forth, and coiling himself about the bottle inhales the perfume of the tincture till he soon becomes stupid, when they return to the house and despatch him with ease.

They attribute to small doses of this tincture the power of producing a magnetic sleep, and use these means to recover stolen property by administration of it to certain persons who are mediums for the business.

The Indians of Chocó consider this plant a specific antidote for snake-bites, and particularly for that of the much-dreaded Verrugosa (Acrochordon Chocoe).

The plant called Solobasta, in the table of antidotes, is another of the Aristolochias, which I have classified as the

---

* All the houses have roofs of straw or reeds.

A. grandiflora Colombiana; it bears a flower 18″ to 20″ in diameter, of strange and rare beauty. This has the same peculiar shape as that of the A. Colombiana, but on a much larger scale, and the infundibuliform calyx has its periphery prolonged on one side into two contiguous wedge-shaped ribbon-like appendages, of two and one half or three feet in length, the lower half of which are not more than a line in width.

A tincture of the root alone, or of the root, leaves, and flower, or of the flower alone (when it has just opened for the first time) is used by the Curers in cases of bites by many different kinds of serpents.

THE SECRET METHODS OF CURE PRACTICED BY THE MOST CELEBRATED CURERS.

FIRST METHOD.

Try the patient's pulse; should it be normal, give *Valdivia* with *Mantuo* and pulverized *alligators' teeth*, and for a beverage water of *Azota caballo* (a plant). If the pulse is high and strong, and the pulsations are noticeable in the jugular veins, or the patient spits a frothy, sanguinolent saliva, and if before the bite he indulged in any carnal act, to the above remedies add the *Azota caballo* for internal administration, and for a beverage, river water.

If the pulse is feverish, and the aforementioned symptoms are absent, it is a *tabardillo* (brain fever) effect of the bite. Should the patient bleed from any of the pores or natural outlets of the body, let the blood flow if it has a dark color; and when it begins to turn a bright red administer *Valdivia* with *powdered alligators' teeth*,* and to stanch the blood a cup

---

* Many Curers say that a snake never bites a person while he holds an alligator's tusk in his hand.

of the following preparation: Make a decoction of *Culantrillo de pozo* boiled alone, cool it, and when cool add a lump of refined sugar. Should this not stanch the blood, *repeat the dose* after an interval of eight hours, and give *a third dose* after a like interval if the second should not produce the desired effect. In the last case give for a beverage *river water*.

Should the blood still flow from some tooth, or from the mouth, make a decoction of *Almaciga*, *Cloves*, and *Cinnamon in white wine;* apply it to the bitten part as a fomentation, and give a teaspoonful of the same internally. This will stanch the blood.

In every case of snake-bite open the wound with a lancet or knife, from fang-wound to fang-wound, in the form of a cross; rub the cut with two fingers till the blood flows from it, and when this ceases, apply to it a plaster composed of *powdered tobacco, garlic,* and *valdivia,* covering it over with a layer of beeswax. Rub the bitten member above the bite with the same composition of the plaster and rum; should these parts be swollen, steam them as follows: Make a decoction of *Alconcito* (Arist. Colombiana), add a few *bitter orange leaves;* the decoction must be made in an earthenware pot never used before. When it has boiled sufficiently, tie a plantain leaf over its mouth, and cut a short slit in the centre of the leaf, placing the arm or leg over the slit, and covering it with a cloth so as to exclude the air, and so that the swollen part may be exposed to the hot vapor.

Repeat the steaming operation three times, and if the swelling should not have diminished after the third time, although the patient may have perspired freely, cover the swollen parts with *taz-taz* leaves, heated in the sun, and then again over live coals, applying fresh leaves every time they become saturated with the perspiration. In case the *taz-taz* cannot be procured, use leaves of the *Abraza palo* in the same manner, and continue the application till the swelling diminishes or is entirely reduced.

Should the patient be attacked with suffocation or pain in the stomach or abdomen, he has eaten something bitter or acid, or fresh pork; give immediately *Fruta de burro*, cooked till it has a bitter taste; this will produce vomiting.

Should the wound form an ulcer or incurable sore, wash it with *verbena* (water), of the kind called *Rabo de alacran* (Scorpion's tail), applying a leaf of the same to the sore; and finish healing it with pulverized *Cadillo seco*, of the kind called *Amor seco* (dry love) (platonic love).

Should the snake be a water-snake, and the wound form a running sore, wash it with a decoction of *Romero* (*Ledum palustre*) water, and apply to it the powdered leaf of the same, having first calcined or burnt it. Continue this cure till a white or black scab forms on the ulcer, and to complete the cure, use an ointment composed of *clean wax, olive oil tallow,* and *cardenillo*.

Should the patient suffer constipation, apply enemas of decoction of *bicho, aniseed,* and *salt*.

For poisonous wounds made by thorns, poisoned arrows, &c., &c., *tincture of camphor* applied on cloths till the parts turn red is the best remedy. And should the wounded parts yield a sanguinolent liquid, apply to them externally the same powder recommended for sores formed from bites by water-snakes.

Diet must be rigid. Absolutely prohibited are eggs, pork, any alcoholic drink, any acid; coitus; and no pregnant woman, or any woman with her menstrual flow present, must approach or remain near the patient.

SECOND METHOD.

When the bite is that of a *Mapana*, the pulse is slower than usual; when it is that of a *Rattlesnake* (*Cascabel*), it is much more accelerated; and when it is that of a *Coral*, it is hardly perceptible.

Should the patient have a flow of blood from the ears, nose or mouth, administer a thick *mazamorra de almidon* (a kind of hasty pudding), to which add as much powdered *Estancadera* as can be taken between three fingers, and also a lump of refined sugar. Should the flow of blood not cease, administer a beverage of juice of *la cepa de platano guineo* (the stalk of the banana) sweetened with sugar.

If the blood is black or dark-colored, allow it to flow till the color changes to a bright red, then administer remedies to stanch it. Redness of the white of the eye indicates *Tabardillo;* for this, administer water which has been boiled and cooled, sweetened with refined sugar; give a standing bath of water as hot as can be borne, from head to feet. After the bath proceed with the cure.

Should there be any swelling in the bitten part, make a decoction of three entire plants of *Escobilla menuda;* with this bathe the part, making it as hot as can be borne, and cover the swollen surface with leaves of the *Abraza palo;* fifteen minutes afterwards raise the leaves and clean off the perspiration, when a yellow liquid will be observed issuing from the wound; this is the poison.

To cure the remaining ulcer, use scrapings of the heart of a tree called *Palo de vaca*. (cow tree), and if this should not cause it to heal, use powdered *cedron nut.*

For Rattlesnake-bites, take *Carnestolenda* roots, from the side towards the rising sun; pulverize and put the powder in ten ounces of water, administer half the quantity, and cause the patient to snuff a small quantity up the nostrils.

Should there be a flow of blood from any pore or conduit of the body, give a small quantity of water, boiled and sweetened with refined sugar; this at 11 A.M.

If this should not stanch the flow of blood give half a gill (un tanto) of juice of *Cepa de guineo* sweetened with sugar. Should the patient be uneasy or restless, grind up a *yuca*, add to the mass a little sugar, shave a spot clean on the crown of the head, and apply the pulp to the spot. To *Gabaro de carnestolenda* add a little water, and with this rub well the bitten limb, binding it afterwards on the bitten part with a strip of cotton cloth.

For a *Patoquilla*-bite, first administer a beverage of lukewarm water sweetened with refined sugar, and afterwards a decoction or tincture of *Capitana* and *Jenerala* in two or three ounces of water; rub the bitten limb with *Capitana*, and with a lancet make a cut from fang-wound to fang-wound, and another deep one crosswise, and with the lancet scrape off the blood that may flow from the wound, putting upon the latter powders of the fruit called *Ojo de buci* (ox's eye). Should there be any headache, give the patient a standing bath in running water (brook or river) of ten minutes' duration.

If on account of any pregnant woman, or woman with her monthly flow, having passed near the patient, there should be an aggravation of the pains, give the bath as indicated in the preceding paragraph.

For a Mapana-bite administer *Capitana*, or *Capitana jenerala*, or *Jenerala* in a dose with two and one-half ounces of water. Scarify the bitten part from fang-wound to fang-wound, and scrape it till the red blood flows, then bind upon it powdered *Capitana* mixed with *Estancadera*, but if this cannot be procured use powdered *Chupadera* (a wild yam whose stalk resembles a snake). Rub well the bitten limb with a diluted tincture of *Capitana*.

Should there be any œdema give as hot a bath of the *Escobilla* (three plants) as can be supported, and put leaves of the *Abraza palo* upon the bitten limb, after having heated them at the fire, covering the whole surface of the swollen part with them, and wrapping up the whole limb with a clean cloth. While the swelling continues, change these leaves by renewal every night and morning.

Should there be a flow of blood from any pore or conduit of the body, give as much pulverized *Estancadera* as can be taken up between the thumb and forefinger, in a little water; and if there is any headache give the standing bath.

For a *Coral*-bite, use the remedies recommended for that of a *Patoquilla*; if this is not possible give a standing bath immediately.

Tincture of *Arponcito Alconcito* (*Arist. Colo.*) mixed with tincture of *Capitana* (Arist. grand.), in five ounces of water, is good for a *Coral*-bite.

The *Cruceta* is good for bites of any kind of snakes; it should be taken in two and one-half or three ounces of warm water.

The *Solona* and also *Nicha*, administered in warm water, are good for any kind of bites; the latter can also be applied in the form of a plaster on the wound, the whole to be covered with leaves of the *Abraza palo*.

Should the patient have performed a carnal act before being bitten, beat together three eggs, add to them a little sugar, beat them again, and with the froth anoint the back between the shoulder-blades, giving the balance of the eggs internally before beginning the administration of other remedies.

The *Mapana*, called Bogui dorada (Vip. Lach. os flavus), has the mouth and tip of the tail of a bright yellow; its bite may be cured with the same remedies used for a Rattlesnake-bite.

Diet.—Keep out of the sun and night dew. Aliments prohibited: eggs, pork, ripe roast plantain, bananas, aguacates (alligator pears), cheese; nothing acid; must not go barefooted; must on no account allow another Curer "to lay hands upon" him; he must not allow any pregnant or menstruating woman to come near him; but a maid, or a woman not menstruating, *who is entirely exempt from all carnal commerce,* may wait upon him.

THIRD METHOD.

A third method is very similar in the treatment, and concludes as follows:

For the bite of the *Sierpe* (a Python) open a hen *alive*, and put one-half of it upon the wounded part, binding it fast with a bandage.

Diet.—The first day of the cure give the patient fowl; when better, beef and salt fish may be permitted. *Absolutely prohibited:* Fish, eggs, banana, aguacates, cheese, anything acid or flatulent; great care must be taken not to go barefooted; and *on no account to step on fowls' excrement* (this is mortal!); total abstinence from any carnal act; and he must not allow any woman, pregnant or menstruating, to come near him while being cured.

Other methods might be given, but they are quite similar to those already noted, which are curiosities in their way, insomuch as that they have never before been published.

Of the Galls, and their Mode of Preparation.

It must always be borne in mind that the gall of a snake has its virtues most fully developed shortly after the skin has been cast, and when its poison is most venomous, provided, however, that the reptile has not eaten any food in the meantime, in which case the gall-bladder will be found nearly or quite empty.

Many experiments with this substance combined with alcohol, in widely different proportions of each ingredient, have led me to adopt the following as the method of preparation which has proven itself most efficacious in a great number of cases:

Proportion: *one drop* of pure gall to *ten drops* of as pure alcohol, or high wines, or spirits of wine, as can be procured.

The mixture must be thoroughly shaken, and allowed to stand for a couple of days, when a lead-colored sediment will have deposited itself; the supernatant liquid can be poured off carefully into a perfectly clean, new vial, using a bit of sponge in the neck of a small funnel to filter it, when it is ready for use.

Never mix galls of different species.

Have the vials distinctly labelled.

The galls of the Centipede, Tarantula, Scorpion, and Cricket can be separated by bruising the insect in a small quantity of alcohol, allowing it to stand for twenty-four hours, and filtering as above mentioned. Each one of these may be used for the bite of the insect, that of the Cricket being a common remedy in some places for a suspension of the urine.

I have used these preparations of snakes' galls topically in all cases of cuts, flesh wounds, or abrasions, either by poisonous thorns or otherwise, and also in cases of poisoning of the surface of the skin by vegetable poisons; the cure is always prompt, and no inflammation supervenes.

It has been said that the gall of the alligator applied topically will remove a cataract; many persons suppose that this is the remedy employed by the famous charlatan Perdomo

Neira, who is said to cure cases of cataract in from three to thirty days.

Method of Administration.

For all ordinary cases of bites, five or ten drops of prepared gall (selecting if possible that of *the kind* causing the bite) in half a tumblerful of water, well mixed, to administer a tablespoonful of the mixture every five, ten, fifteen, or twenty minutes, according to the violence of the symptoms, and varying the dose from three to ten drops of the gall, according to the age, sex, condition, and susceptibility of the patient, will afford entire relief.

In fifty cases treated, I have given ten-drop doses of gall in four ounces of water *in two cases* only; all the others have been cured by five-drop doses, continuing the remedy at more prolonged intervals as the symptoms of the action of the poison disappear; and in the two cases cited, where a five-drop dose was given in tablespoonfuls every five minutes without producing relief, the dose was immediately repeated, and entire relief ultimately ensued.

I invariably make a deep cruciform incision in the wound with a lancet, and bathe the limb in water as hot as can be borne, into which I pour a few drops of prepared gall.

When the blood flows a bright red (and not before) a small pellet of cotton or sponge, saturated with the gall (prepared), applied to the wound and secured with a bandage, will stop the flow of blood, unless a large vein or artery is punctured by the fang; in this case cauterization is necessary.

THE POISONS AS REMEDIAL AGENTS.

Dr. Mitchell concludes his remarks upon the action of the venom on the tissues and fluids with the following remarkable sentences:

Analogy between Crotalus Poisoning and the Symptoms of Certain Diseases.

"I am unwilling to leave this unsatisfactory but necessary part of my task without calling attention to the singular likeness between the symptoms and lesions of Crotalus poisoning and those of certain maladies, such as yellow fever.*

"If for a moment we lose sight of the local injection, and regard only the symptoms which follow and the tissue-changes which ensue, the resemblance becomes still more striking.

"In both diseases—for such they are—we have a class of cases in which death seems to occur suddenly and inexplicably, as though caused by an overwhelming dose of poison. In both diseases these cases are marked by symptoms of profound prostration, and in both the post-mortem revelations fail to explain the death. I have spoken, as an example, of yellow fever, but similar instances are not wanting in cholera, typhoid and typhus fevers, and in scarlatina.

"A second class of cases, both of Crotalus poisoning and of yellow fever, survive the first shock of the malady, and then begin to exhibit the train of symptoms which terminates in more or less degradation of the character of the blood, varying remarkably among themselves, exhibiting, as it were, preferences for this or that organ; all these maladies agree in the destruction of the fibrin of the blood, which these fatal cases frequently exhibit. In yellow fever the likeness to the venom poisoning is most distinctly preserved as we trace the symptoms of both diseases to the point where the diffluent blood leaks out into the mucous and serous cavities. The yellowness

---

* This analogy has been noted by L. S. Mitchell, by Magendie, and by Gaspard, who has also called attention to the resemblance between ordinary putrefactive poisoning, such as arises from injection into the blood of decayed animal substances, and the poisoning by venom.

which characterizes many yellow fever cases I do not find described as a current symptom of the venom malady, but it is often mentioned as one of the accompaniments of the period of the recovery from the bite.* It is, indeed, most probable that if small and repeated doses of venom were introduced at intervals into the body of an animal, a disease might be produced even more resembling the malady in question.

"In the parallel thus drawn I have given the broad outlines of resemblance, nor was it to be expected that the minor details would be alike. From a general and philosophic point of view, this similarity is sufficiently striking to make me hope that the complete control of one septic poison for experimental use may enable us in future to throw new light on those septic poisons of disease, of whose composition we know nothing, and whose very means of entering the body they destroy is as yet a mystery."

Pathogenetic tables, showing the effects of the different poisons upon the system when taken into the stomach, are as yet quite incomplete, and for this reason only synopses of some of them are given. It is the Author's intention to collect all reliable data together, and thus make this study more complete, so that what any one desires to know of every branch of this subject may be found in another edition of the present work.

---

* Jaundice, occasionally observed in France, as an early symptom of viper-bite, has been usually regarded as jaundice of fear; a cause which certainly cannot be invoked to account for the icterus seen in the last stages of the malady caused by the venom.

Crotalus Horridus.

A résumé of symptoms from provings by Dr. Mure is:
Mental prostration.
Debility of intellectual faculties.
Visions; extraordinary hallucinations; almost lunacy
Illusions of the sense of hearing; deafness.
Sanguineous congestions; choking sensations; cramps.
Pains in the limbs; skin covered with pustules.
Coldness, with increase of the pulse.
Derangement of the menstrual flow.
Slight symptoms of inflammation of the liver.
The above are noted in their chronological order.

Lachesis Trigonocephalus.

The symptoms furnished by Dr. C. Hering and Dr. Jahr are:

Synopsis.

Tension in the muscles as though they were too short.
Sharp, dragging rheumatic pains in the limbs.
Intermittent and periodical pains; sufferings accompanied by a fear of suffocation.
Aggravation and renewal of pains after sleep, or some hours after a meal.
Great physical and mental debility; rapid diminution of strength.
Fainting fits accompanied by dyspnœa, nausea, cold sweats, vertigo, and pallor in the face.
Accesses of asphyxia and syncope.
Attacks like epileptic fits; convulsions with screams and movements of the limbs; before the attacks, feet cold, eructations, pallor of the face, and vertigo.

Hemorrhages and extravasation of blood in different organs.

ELAPS CORALLINUS.

*Symptoms from Provings made for Dr. B. Mure.**

Dr. Mure remarks:

"The number of symptoms noted is not considerable, but I believe I can respond for their exactitude. Almost all of them have been developed in different persons; others have been already confirmed in cases of cures effected, as, for example, feeling of oppression upon ascending stairs, vesicular eruption in the feet, and deafness.

"This last symptom is of the greatest importance, as it corresponds to a disease which, until the present time (1847), has shown itself exceedingly rebellious to the homœopathic treatment.

"Affections of the lungs will probably find a valuable medicinal agent in the *Elaps corallinus*.

"Hæmoptysis and the state of the digestive canal, are special indications in the course of the second period of pulmonary phthisis; next follow mental disturbances and cutaneous eruptions.

"A special action on the organs of the right side, paralysis, lancinating pains ensue; and it appears to me that the gyratory movements, lack of oscillation, fall of the epidermis, and certain moral symptoms, are analogous to the nature of the reptile itself.

"Undoubtedly marked analogies do exist between the symptoms developed by this poison, and those by that of the *Lachesis trigonocephalus*, but there are still sufficiently numerous points of difference to prove that these animal poisons

---

* Pathogenesie Bresilienne.

do not have an almost identical action, as many persons have affirmed; and that the *Crotalus horridus*, for example, is the almost perfect antidote or succedaneum of the *Lachesis trigonocephalus*. I am well convinced of the contrary, and my belief is *that serpent-poisons, well proven, afford in themselves alone the most rapid and surest means with which to combat all known diseases.* Each epoch has doubtless a recourse of therapeutic means, which are most specific, because most homœopathic, to the general character of epidemic or reigning diseases. So that when the human kind shall be partially purified of the miasms that attack the race, the simple flowers of the fields will afford a superabundant store to skilled physicians from which to take agents with which to combat slight indispositions; but we, who are the unfortunate heirs to the chronic miasms of all ages, we who are surcharged with the virus of leprosy, scrofula, syphilis, and a thousand other detestable chancres, must seek in the most powerful medicinal agents resources proportionate to the intensity of our sufferings."

Amphisbœna Vermicularis.*

Fainting fits; feels downcast.

Great pain in the whole spinal column, augmented by motion.

Rose-colored miliary eruptions. Sleeplessness.

Feels a great weight in the forehead and parietal bones.

Intense cephalalgia, vertigo, inclination to fall towards the right side.

Sight affected, tremors in the orbital muscles.

Dull, heavy pains in the bone of the right lower jaw repeated several times. Pains in all the teeth.

---

* Pathogenesie Bresilienne.

Painful deglutition, swelling of the tonsils.

Deglutition impossible; chills; pains in epigastrium.

Constipation. The eruption extends over the breast, elbows, and back, and has an itching sensation, worse in the morning, but relieved at night.

Swelling of the arm with great pain.

Cramps in the left leg accompanied by insensibility of the same.

Aristolochia Milhomens (*Mure*).

Aristolochia grandiflora (*Gom.*). Aristolochia symbifera (*Mark*).

(An Antidote used in Brazil for the Bite of the Rattlesnake.)

Pathogenesy.

Potency used, 5th cent.

Great agitation during sleep.

He dreams that he cannot drink, walk, or move himself.

Sensation of pulsations in right frontal protuberance.

Pasty taste in mouth upon rising.

5. Thirst.

Pain in the right groin.

Sensation of numbness in left leg.

Sensation of numbness in lower part of the calf of same leg.

Rumbling sounds in the stomach and intestines.

10. Pricking sensation in the whole of left leg.

Sensation of heaviness in the head.

Great thirst, accompanied with a bitter taste in the mouth.

Loss of appetite.

Redness and swelling of the left leg.

15. 3.30 P. M. A pricking, itchy sensation in the hypothenary protuberance of the right hand.

Sensation as of a weight on the crown of the head.
A pricking, itchy sensation in the right testicle.
Pricky sensation in the right thigh.
Sensation as though pricked by a pin in the lower part of the left leg.
20. Pricking sensation in the internal part of the left thigh.
Sensation as though pricked with a pin in different parts of the body.
7 P.M. Repetition of symptom No. 15.
Feeling of heaviness in the cerebellum.
Sensation as of being pricked in the heel.
25. 8 P.M. Itching sensation in the right side of left foot.
Pricking and itchy sensation in the skin of the prepuce.
Cramplike pain on the internal edge of right foot.
Contusive pain in left pectoral muscle, which is exceedingly sensitive to the touch at night.

Second Day.

Restlessness during sleep.
30. He dreams that he sees a sheep and a dog, dressed in red; a man, accompanied by many other individuals, holds the dog by the back suspended in the air, and the dog holds the sheep suspended by the middle of the back while the latter moves his head with an oscillatory movement.
This is followed by an erotic dream, with an involuntary flow of semen.
Feels a pain under the shoulder blade as if from a blow.
Dull pain on the internal edge of left foot.
2 P.M. Sensation of inquietude; afterwards a pricky, itchy sensation in the muscles of the body.

35. Contusive pain in the left knee.
    7 P.M. Lancinating pains in the anterior part of the external edge of the left foot.
    Sensation of repletion in the stomach.
    Sharp pain at one point in the right thigh.
    Blue-black œdema of the leg in the morning; the leg becomes more inflamed with exercise, and is much darker and discolored at night.
40. Lack of appetite.
    The whole of the leg is covered with dark, irregular-shaped patches or spots formed by the extravasation of the blood.
    Frequent voiding of urine.
    Burning sensation in the head.
    Continued thirst with a bitter taste in the mouth.
45. Excoriation of the lips and gums.
    Total loss of appetite.
    Pain as of excoriation in the left leg, which afterwards passes to the internal edge of the right foot, where it is felt more acutely.

Third Day.

An almost unbearable pain in the lower part of the heart that checks respiration; this is felt during the night.
Temporal regions exceedingly painful to the touch during the whole day.
50. Rigidness of the leg with impossibility to remain standing, which lasts several minutes.
    Sharp pain in one point between the shoulders.
    Dull pain in the lower lumbar and hypogastric regions.
    Burning pains in the anus.
    Cracks in the lips and gums as on the preceding day.

55. Pricky, itchy sensation in the flexor of right arm.
Cramp-like pain in the left tendo-Achillis.
Sensation of numbness around the edges of the feet.
3 P.M. Contusive pains in the left knee.
3.30 P.M. Sensations of being pricked with a pin in the lower part of the right leg, and on the internal edge of the foot.
60. 4.30 P.M. The whole of the left arm is painful to the touch.
Pricking pains in the lower part of the left knee.
8 P.M. Itching sensation in the articulation of the first phalanx of the little finger.

Fourth Day.

Pain in the back of the left index-finger.
Colic, followed by a stool of soft matter, and diarrhœic stools the next morning.
65. Sensation of uneasiness as if a tumor was being formed on the inner side of the right leg above the knee; this symptom was felt in the P. M.

Fifth Day.

Heaviness in the lumbar region.
Sensation as if the skin of the lower part of the leg was sliding down towards the ankle, with a frequent inclination to draw it up again.
Itching sensation in the lower part of the right leg.
Pricking, itchy sensation on the internal surface of the right leg.
70. A strange sensation in the lower part of the tendo-Achillis.
Easy stool.

Sixth Day.

Pricking, itchy sensation on the internal surface of the right ankle.

Itchy sensation in the muscles of the left side.

Pain on the internal surface of the leg just above the right ankle.

75. Unpleasant dreams.

Seventh Day.

Restlessness; after awakening in the morning impossibility to go to sleep again.

He feels for several hours as if something was troubling him about his ankles.

3 P.M. The pain increases and becomes contusive.

Ankles appear swollen.

80. Acute pain in the sacro-lumbar region.

Pain in the right hypochondria.

Eighth Day.

Pain in the scrobiculum.

The pains in the leg continue.

Fixed pain on the internal surface of the left ankle-joint.

85. Burning pain on the internal part of the right leg at night.

Acute lancinating pains in the head at night.

Ninth Day.

Acute lancinating pains in left side of head at night.

Tenth Day.

A deep, lancinating pain in back part of head.

The fore part of the left leg is painful to the touch.

90. Lancinating pains in the cerebellum.

Many symptoms are common with those developed by Crotalus horridus, and thus explain its action as an antidote.

ARISTOLOCHIA COLOMBIANA (*nobis*).

Great agitation and restlessness at night.

Dreams of inability to do certain ordinary things.

Unpleasant taste in the mouth.

Sensation of pulsations in the orbits of the eyes.

5. Thirst constant during the day.

Pains in the joints of the legs.

Pains in the joints are fugitive.

Sensation of numbness, sometimes in one leg and sometimes in the other.

Rumbling in bowels, sometimes with and at times without colicky pains.

10. Sensation in the legs as of being stung by nettles. Both legs red and swollen, particularly below the knee.

Continued thirst *with a taste of the plant in the mouth.*

Agrypnia. Anorexia.

Sensation as of being pricked by pins in right leg.

15. Same sensation felt afterwards in the other leg.

Cramps in the feet.

Contusive pain in the chest (left side).

Has strange fantastic dreams.

Rheumatic pains under the shoulder-blades.

20. Rheumatic pain in left instep.

Restlessness, repetition of symptoms No. 10.

Legs are much swollen and discolored.

Lack of appetite continues.
Surface of the leg discolored in spots. Legs much swollen, with painful varices in pregnant females.
25. Increased flow of urine.
Continued thirst, *with a taste of the plant in the mouth.*
Loss of appetite continues.
Fugitive pains below both knees.
Pains of a rheumatic nature in the shoulders.
30. Rheumatic pain in the lumbar region.
Cutting pain in the rectum after a stool.
Cramp-like pain in the left instep.
Rheumatic pain in left knee of short duration.
Fugitive pains in the arms and shoulders.
35. Pains in left knee continue.
Sensation as of a boil being formed on the thigh of the right leg, which is very painful to the touch.
Itching sensation in the skin below the same knee.
Rheumatic pains in both ankles.
Sleeplessness.
40. Pains in lower limbs continue, and become almost unbearable, extending up to the hips and lumbar region.
Acute pains in the crown of the head.
Lancinating pains in the cerebellum.
43. Paralysis of the tongue.

I can respond for the exactitude of nearly all the above symptoms, as they have disappeared in different patients under the administration of doses of the O, 3d cent. and 6th cent. potencies. It has proved successful (at the 6th cent.) in two cases of suppressed menstruation with rheumatic pains in the lower extremities. In a great number of cases of Scorpion-bites, in which paralysis of the tongue is always present, it has invariably afforded entire relief *in less than five minutes.*

ADDITIONAL CLINICAL NOTES ON THE ACTION OF THE POISONS FROM DIFFERENT SOURCES.

The employment of *Lachesis* in gangrene is shown to be exceedingly efficacious, by three cases of cures cited by Dr. Hiatt,* in which its action was quick and specific.

Its action is more strongly marked when the part assumes a purple or lardaceous hue, and emits an offensive odor. Its efficacy in warding off traumatic tetanus and gangrene, in cases of gunshot wounds, has been proven in a great many cases; possibly a single case cannot be cited in which it has failed to produce beneficial effects. Kuhn gives a case of a soldier bitten by this serpent, where the hand and arm were much swollen, and covered with gangrenous blisters. The wound gangrened, necessitating subsequent amputation to save the patient's life. A notable fact in this case was that the urinary and fecal discharges were suppressed for *seven days.*

In cases of yellow fever many symptoms are found like those under *Apis mell., Crot. horr., Lachesis,* and *Elaps corall.,* viz.: General tremors; extreme lassitude; jerking and convulsions of the muscles of the limbs; delirium; sudden starting up from a sound sleep; frenzy; debility and numbness of the muscles of the limbs; fainting fits; loss of consciousness, and sudden sinking of the vital powers.

In this connection I desire to call attention to the action of the poison *Vip. Lachesis os flavus,* as its similarity with the symptoms which mark the last stage of yellow fever cases *is*

---

* An Inaugural, North American Journal of Homœopathy, May, 1865.

still more striking even than what has already been noticed in cases of injection of *Crotalus* venom.

Thus far the poisons which have been proven take the following rank in regard to their action in decomposing the blood, viz.:

    1. Crotalus.
    2. Vipera redi, Elaps corall.
    3. Lachesis trig., L. os flavus.
    4. Naja tripudians.
    5. Naja haje.

Lachesis shows a remarkable power in allaying the cough in phthisis and pulmonary abscesses, and *Crotalus* in phthisis alone.

"Facts are stubborn things," but still, in opposition to the repeated asseverations of the most highly respectable authorities in some matters to the contrary notwithstanding, they do most certainly and positively prove that *all* the pathogenetic effects of any medicinal agent *are only developed through the entire scale of potencies.* Dr. Fincke's 100,000th potency of Lachesis furnishes proofs of this in cases cured, and in the development of *new symptoms* which had not shown themselves in the lower potencies. Some other and more astounding facts are in reserve for a future occasion, when the "potency question" shall have been more fully studied.

In relation to the neurotic symptoms the order is at present as follows, viz.:

    1. Lachesis trig.
    2. Naja trip.
    3. Crotalus.
    4. Vipera redi.

But when the substance can be furnished for provings, *Lachesis niger* will head the list.

The sanguineous sphere is most affected by Apis mell. and Theridion curassavicum, while the reproductive sphere is most powerfully affected by Bufo sahytiensis and Tarantella. One of the most important indications for *Bufo s.*, and which I have never seen noticed in any work, is *Impotency!*

The "wise women" or "midwives" in South America hang up a frog by the hind legs till a greenish saliva drools from its mouth; two or three drops of this is caught in a spoon, and furnished (for a consideration) to some jealous female, who administers it to her husband (unknown to any one but herself and the "old crone") in a plate of soup or food of any kind. This produces a temporary impotency which continues for *two or three months*, and then gradually disappears of itself.

Sometimes the dose given is larger, and the person to whom it is administered is nervous and susceptible; the result is permanent impotency. I have known several of these cases which I had the most powerful motives for attributing to the above causes.

One of the most prominent symptoms of *Elaps corallinus* is *deafness;* and its powerful action on the respiratory organs is similar to *Lachesis.*

Dr. Mure reports having cured several cases of chronic deafness (without any existing organic lesion of the auditory organs) with this substance; and I have administered *Elaps corallinus gall*, 3d cent. and 6th cent., in two similar cases with perfect success. One case, of a man who had been deaf in both ears for ten or twelve years, was worthy of note.

After putting the patient on diet for a week, prohibiting the use of tea, coffee, and tobacco, a dose of the gall 6th, was given, and a powder of Sacch. lact. daily for a week, then semi-weekly for three weeks more. For 32 days after the ingestion of the remedy no apparent effect had been produced but upon rising on the morning of the 33d day he experienced, to use his own expression, "something like the bursting of a fish-bladder," followed by considerable distinctness of hearing in one ear first, and then in the other; on the 45th day after ingestion he reported being able to hear as distinctly as he ever could in his life.

This case affords one proof at least of my assertion that the action of these galls *pathogenetically* is a perfect *similimum* of the *toxical* action of their congenital poisons. Let me give another in this connection which may not be found devoid of interest.

In the town of Juan de Acosta, nine leagues from Barranquilla, in the State of Bolivar,* in 1870, a countryman was bitten by a *Vipera Lachesis bufocephalus*, and brought into the town from his little plantation, together with the snake that caused the bite, which he had killed.

Some one of the bystanders proposed to give him the gall of the reptile, adding that this was the remedy which I used to cure snake-bites. He assented. The gall-bladder was taken from the snake and emptied into three or four ounces of common rum (the quantity of pure gall could not have been less than from 100 to 150 drops), which was given in a single dose. All the pains ceased as if by magic, and in half an hour the bitten man said he felt perfectly recovered. An hour elapsed; there was a sudden flow of blood from the nose, followed by acute darting pains in the bitten limb, and a subsequent development of all the toxical symptoms of the poison, terminating in death at the lapse of six hours.

An explanation of the action of the gall in this case was undoubtedly the extinction of all the symptoms of the action of the poison in a short time, and then that of the gall became

---

* United States of Colombia.

a medicinal aggravation, which, owing to the massiveness of the dose, became so enormously powerful as to extinguish the vital force.

Those persons who knew of the case considered that the gall was poison, and quite as deadly as the venom of the serpent itself.

A few weeks later, a snake of the same kind bit another man at or near the identical place where the first one had been bitten. A gentleman for whom the bitten man was laboring had a small vial of gall of the same kind of snake, prepared by myself, which was administered according to the directions given on page 209. After the lapse of two hours every symptom of the poisoning had disappeared, and the man was able to work the ensuing day. No subsequent relapse or return of any of the pains took place. I leave the reader to draw his own conclusions as to the causes of the result in both cases.

Dr. Baruch has effected cures of phthisis florida with *Theridion curassav.*, administering this remedy while the disease was yet incipient.

Dr. Lilienthal's inference with respect to animal poisons is that they all act alike, producing adynamia with sanguineous or nervous depression; and consequently remove these states when homœopathically indicated.

The Crotalus poison, says Dr. Lilienthal, produces always, sometimes in a few minutes, painful swelling of the bitten part, ecchymosis; bluish-gray color of the adjacent cutaneous tissue, and gangrene, with hemorrhages from nearly all the orifices of the body; and through this decomposition a depression of the nerve-centres, showing its action by nervous twitchings and convulsions, delirium, syncope, exhaustion, and death, from paralysis of the spinal nerves.

The *viper's* (*torva* and *redi*) bite also produces gangrene (ultimately) and hemorrhages from the different orifices. Cases are recorded where the bitten persons fell down immediately after the bite, and at a later period paralytic symptoms always developed themselves.

Post mortems have shown that the oppression in the chest with its torturing anguish (so distinctly characteristic of *Lachesis niger*, *Lachesis trigonocephalus*, and *Elaps corallinus*), is very frequently caused by the extension of the gangrene to the lungs and liver; although death may supervene (as by *Crotalus*), by the depression of the nerve force; so necessary is this latter to the sustenance of the functions of life.

In *Lachesis* we see the nerve-centres attacked first; the man falls as if struck by lightning! Unconsciousness follows (at least in some cases); the sympatheticus and vagus are attacked, thus disturbing the whole machinery of life, and causing the decomposition of the blood, thus producing gangrene, which is but a consequence of this lack of vitality.

The *Naja tripudians* affects especially the pneumogastricus, and the neurotic symptoms predominate over the hæmatic. It produces instant paralysis, destroying or annihilating the source of nerve-force; affecting first the sensory, and later the motor nerves. Hughes states that in the case of the keeper killed in the Zoological Gardens in London by the bite of a Naja, the air-passages were filled with a frothy mucus, and that death ensued from suppression of the respiration. This case affords a proof of the preceding statement of its action.

This indication in the case just cited is a precise counterpart of deaths in many cases in verminous or helminthic affections. Post mortems invariably show the air-passages filled with a frothy mucus.

A prominent symptom in many cases of Cobra-bites is

vomiting a large quantity of black fluid; and in some instances this is followed by a copious evacuation of a thin whitish or brownish liquid, very fetid, and mixed with partially softened fæces. The pupil of the eye seems to be contracted till death ensues, when it expands to its utmost limit.

BIBLIOGRAPHY AND EXCERPTIONS.

Pathogenesic Bresilienne, by Dr. Mure, Paris, 1847.

Thanatophidia, by J. Fayrer, M.D. London, 1872. By J. & A. Churchill.

Buffon's Natural History (Reptilia). Vols. 83, 84, 85.

Wood's Natural History (Reptilia). Vol. 3.

Zoology of South Africa, by Andrew Smith, M.D. London, 1849.

Philosophical Transactions Royal Society, Vol. 79.

Kempfer's Amœnitates Exoticæ.

Gronovius (L. T.) Zoophylacium, No. 96. Leyden, 1787.

Russell's India Serpents. 2 vols., folio. London, 1796, 1801.

Blasius Merrem Beytrarge zur naturgeschichte der Amphibien. Dinsbourg and Leipsic, 1799. 2 vols.

Seba (d'Albert) Rerum Nat. Amst. 1734, folio.

Amphibia (G. Schneider), 2 vols. Jena, 1801.

Asiatic Researches (John Williams), 2 vols. Calcutta, 1790.

Zoology (George Shaw), vol. 3. London, 1800–1826.

Medical Remembrancer (E. B. L. Shaw). 4th Edition. London, 1856.

Dr. Mead's Account of the Viper. Appendix by Dr. Mitchell.

Traité sur le venin du vipere. By Abbe Fontana. Translated by J. Skinner. London, 1787.

Lacepede (Gallus), 1st vol.
Ptolemao Philadelpho. Lib. 16. Cap. 39.
The Reptiles of British India. By A. C. L. G. Günther. 8vo. London, 1864. Robert Hardwicke, No. 192 Piccadilly. Published by the Ray Society.
An Illustrated Catalogue of all the Snakes of Australia. By Gerard Krefft. Sydney, 1869. 4to. Printed for the Government, by T. Richards.
Medical Zoology by Stephenson.
Bell's British Reptiles.
Darwin's Voyage of the Beagle.
Jones's Natural History.
Smithsonian Contributions to Knowledge. Researches upon the Venom of the Rattlesnake. By S. Weir Mitchell, M.D. D. Appleton & Co.: New York.
Crotalus Horridus and Yellow Fever. By C. Neidhard, M.D. William Radde (now Boericke & Tafel): New York, 1868. Second Edition.
Notes on Kaferistan, by Captain Raverty.
Wise's History of Medicine among the Asiatics.
Schneider; Amphibia. Jena, 1799.
Geography of the United States of Colombia. Edited by F. Perez. 2 vols. Bogota, 1862.
Prince Lucien Bonaparte's Analysis of Viper Venom. 1843. Gaz. Tosc delle se Medicofis.
Catalogue of North American Reptiles; Smithsonian Institution. By S. F. Baird and C. Girard. Washington, D. C., 1853.
Rymer Jones, Anatomy and Physiology.
I. L. Soubeiran, De la Vipere. Paris, 1855.
Owen, On the Skeleton and Teeth. Philadelphia, 1854.
Claude Bernard. Leçons sur les effets des substances toxiques, etc. Paris, 1857.

Robert Boyle. De Utilitate, &c. London (Lindiviæ). 1 vol. 4to. 1692.

David Brainerd, M.D. On the Nature and Cure of Bites of Serpents, &c. Smithsonian Reports, 1854.

Cantor. London Zoological Transactions. II.

Jean Louis Chabert. Du Huaco (Guaco) et ses vertus medicinales. 8vo. 1853.

John Davy on Snake-stones. Asiatic Researches. Vol. XIII. 4to.

Dumeril and Bibron. Erpétologie Générale, VI et VII. Paris, 1844.

Geoffroy and Humauld. Mémoire de l'Académie des Sciences. Paris, 1737.

Herran. Journal de Pharmacie. 3d Series, XVIII. Paris, 1850.

Kalm. Travels in America. 1753.

Koster. Voyage au Brésil. II, p. 247.

Nicholas B. L. Manzini. Histoire de l'inoculation, &c. Paris, 8vo. 1858.

Dr. George B. Halford. A paper read before the Victoria Society of Melbourne, Australia, June, 1870.

Orfila. Traité de Toxicologie. Paris, 1852.

INDEX.

Adder, Puff, of Africa, 97
    Death, of Australia, 98
Africa, South, Vipera in, 100
Ammonia, externally, and injected, useless as an antidote to snake-poisons, 139, 143
Analogy between Crotalus poisoning and certain diseases, 211
Antidotes of serpent-venom, used by Curers, 186
    list of 41 substances used by Curers, 188
    galls of serpents, 192, 199, 209
    Aristolochia Colombia, 193
        directions for using, 194
    serpentaria Virginia, 196
    cedron, simaba cedron, 196
    Valdivia nut, 196
    guaco, 196, 197
    alcohol, remarkable case, 198
    Dr. Mitchell's list of, (5), 199
    cocksflower, 45, 200
    Solobasta aristolochia grandiflora Colombia, 201
Antioquia, United States of Colombia, Vipera in, 102
Aristolochia milhomens, provings of, 216–221
    Colombiana, provings of, 221–222
Australia, Vipera in, 101
Author's experience, prefatory notice of, 9

Baruch, Dr., cures phthisis florida with theridion curassav., 227
Bernard, M., experiments on effects of poisons, 125
Bibliography, list of works referred to, 230–232
Blood, human, analysis of (figure), 116–119
    effect of snake-poison in, 144–148
Boa constrictor, 36

Boericke, Dr., his specimen of Lachesis trigonocephalus described, 94
Bolivar, United States of Colombia, Vipera in, 102
Bora, 43
Boyaca, United States of Colombia, Vipera in, 104
Buffon's classification of Ophidians, 13
Bufo s. causes and cures impotency, 224

Carbolic acid, action of, on Cobra and on frog, 164–171
Cauca, United States of Colombia, Vipera in, 104
Caquetta, United States of Colombia, Vipera in, 105
Classification of serpents, Günther's, 21
Cobra, taming of, 52
    changeable color of, 76
    vomiting of blood caused by Cobra-bites, 228–229
Copperhead, 63
Coral snake, 64
    curious mode of cleansing poison-fangs of, 65
    varieties of, 67
    sexual commerce of, 67
Cotton-mouth, water moccasin, 98
Crotalus, fetid smell of, whence the name, 86, 100
    "bites high," 87
    fasciatus, banded rattlesnake, 81
    poisonous action of, Dr. Lilienthal, 227
Crotalidæ, thirty-nine varieties, 75–88
Cundinamarca, United States of Colombia, Vipera in, 105
Curers, snake, superstition of, 96
    secret methods of, 202–208
    recognize the snake by the bite, 97

Deafness, chronic, cured with Elaps corallinus gall, 225
Death-adder of Australia, 98
Deaths, annual, from snake-bites in India, 134

Elaps corallinus poison, effects of, 108
Elapidæ naja, forty varieties, 48–55
    ophiophagus, many varieties, 55–68
Electricity, relation of atmospheric, to venom of serpents, 86, 116
Experiments (124) with serpent-poison on fowls and animals, 161–173
    summary of, 174–175

INDEX. 235

Family groups of serpents, four, 21
                                      eight, 22
Fangs, poison, 47–48
            extracted, 53
            of aquatic snakes, 72
            loss and repair of, 100
            very long in Callophides intestinalis natis, 59
Fayrer's, Dr., report of cases of poison, 153-161
Flame snake, 97
Fontana, Abbe, experiments of, 120
Formula expressive of poison action, 109

Gall, animal, antidote for its poison, 9
    of snake, relation of, to its poison, 136
        an antidote to its poison, 148, 192, 199
        mode of preparation, 208–209
            administration, 210
        fatal effect of 100 drops, 226
Gangrene caused by Viper-poison, 228
Günther's classification of serpents, 21

Halford, Prof. G. B., experiments with hypodermic injections of ammonia, 148
Hering, Dr. C., Lachesis trigonocephalus of, 94
Height above the sea, relation of to serpent-venom, 185
Holmes, Dr., "Elsie Venner," reference to, 192
Horned viper, Cerastes, 98
Hydrophidæ, seven species, 68–74

Impotency caused and cured by Bufo s., 224

Kabirájes, "fathers in medicine," their preparation and use of snake-poisons in medicine, 150
Krait, 56, 160

Lachesis-poison, sudden action of, 228
        vipera trigonocephalus, of Hering, 94
Library of British Museum, 6
Ligature, inefficacious in snake-bites, 138
Lilienthal, Dr., on Crotalus poison, 227

Magdalena, United States of Colombia, Vipera in, 106
Menstruation in woman aggravates serpent-poisoning; increases the hemorrhage, 191
    influence of on living serpents, 191
Methods of cure (three), secret of the Curers, 202-208
Mitchell's, Dr. S. W., on rattlesnake and Crotalus-poison, 120-132
    conclusions, 132-134
Mure, Dr., 67
Music, influence of on snakes, 53

Naja tripudians, neurotic action of, 228

Ophidia colubriformes, uropeltidæ, 26
    calamaridæ, 27
    oligodoritidæ, 28
    colubridæ, 29
        Æsculapii, 29
    comalopsidæ, 34
    psammophidæ, 36
    dendrophidæ, 36
    dryiophidæ, 38
    dipsadidæ, 38
    lycodontidæ, 39
    amblycephalidæ, 39
    pythonidæ, 40
    erycidæ, 43
    acrochordidæ, 44
  venenosi, elapidæ, 45, 48, 68
    general description of, 45
    poison-fangs of, 47
    hydrophidæ, 45, 68, 74
  viperiformes, crotalidæ, 45, 75, 88
    viperidæ, 45, 88, 116
    description, general, of, 46
    cobra, 49, 53
    naja ophiophagus, 55
  colubriformes, 22
    typhlopidæ, 22
    tortricidæ, 25
    xenopeltidæ, 25
Ophidians, Buffon's classification of, 13

INDEX.

Phosphorescent properties of Vip. cal. rub., 112
Panama, United States of Colombia, Vipera in, 106
Poison of snakes, nature and analysis of, 119
    Crotalus, 119, 120, 121, 227
    rattlesnake, 120
    Crotalus, effect of on Crotalus, 123
        other vipers, 124
  snakes, mode of experimenting with, 137
  Cobra, experiments with, in London, 139
  snakes, use of as a medicine, in Bengal, 150
  Cobra, action of on cold-blooded animals, 162
    and Bungarus, action of on cold- and warm-blooded animals, 163
    Daboia, and strychnine as antidote, 166
      on horses and other animals, 167
  snakes, 124 experiments with, 161–173
  Cobra, appearance of under a microscope (plate), 177
    in blood of a dog and a fowl (plate), 178
  Vipera daboia Russellii in blood of a dog (plate), 179
Poisons as remedial agents, 210
  of Crotalus horridus, proving of, 213
    Lachesis trigonocephalus, proving of, 213
    Elaps corallinus, proving of, 214
    Amphisbœna vermicularis, proving of, 215
Poison of snakes in blood of fowls (plate), 179, 180
    crystals from (plate), 182
    effects of on muscular fibre (plate), 183
Poisons, action of, clinical notes on, 223
  relative value of as medicines, 224
  cures by, Dr. Mure, 225
Puff adder, of Africa, 97

Rattlesnake, banded, 81
  description of in North and South America, 81, 86
  poison of, a cure for leprosy, 87
    fatal experiment with, 87
Rotter, Podre dora, birri, 99

Santander, United States of Colombia, Vipera in, 107
Secret methods of cure by the Curers, now first published (three), 202–208

"Serpents, fiery," of Bible, introduction, 5
    four family groups of, 21
        eight family groups of, 22
        North American, table of species of, 14–20
Serpent worship, ancient, 11
            symbolism of, 11
Sfigmas, illustration of serpent-poison, 177, 184
Snake, double-headed, 23
    blind, 24
    milk, or house, 29
    thunder, or king, 29
    black, Schwartze schlange, 32
    green, 33
    Fresh water, 34, 68
    grass, or ring, 35
    Cockscomb, 36
    Tree, dendrophidæ, 36
        golden, 38
    whip, coach, emerald, 38
    nocturnal tree, 38
    ground, 39
    blunt-heads, 39
    rock, 40
        diamond, 40
        carpet, 40
        black-headed, 40
        children's, 40
        olive-green, 40
        Gilbert's, 40
    aboma, rigid boa of Mexico, 40
    Port Natal, 25 feet long, 40
    common (East Indian), 40
    bora, 31 feet long, 43
    edible (East Indian), 44
    sand, 43
    hood, 52
    wart, 44
    copperhead (cerastes), 61, 62, 63
    coral, 64
    spew, 55
    aquatic, 68

Snake, banded, rattle, 81
    flame, 97
Snake-bites, 65 cases officially reported to East India Government, 153–160
Species of North American serpents, table of, 14–20
        other serpents, 20
Shields on head of snakes, 46

Tanjore pill, composition of, 173
Toxical effects of Elaps corallinus, 108
                Vipera lachesis bufocephalus, 109, 226
                          os flavus, yellow-mouthed, 110, 148
                          niger, 111
                          calamaris venenosus, 111
                          pseudechis major, 111
                          echis striata, 111
                          calamaris venenosus rubrum, 111
                Crotalus cascabella, 112
Theridion curassav., in phthisis florida, Dr. Baruch, 227
Tolima, United States of Colombia, vipera in, 107

Venom-gland, structure of (plate), 184
Viper, lives a year without food or water, 91
    niger, receives her young into her mouth, 94
    horned, cerastes, vicious-looking, 98
Vipers, additional varieties, 100
        21 South African, 101
        27 Australian, 101
Vipera lachesis trigonocephalus (Hering), 94
    lance-headed, 88
    fatality of poison of, 88
Viperidæ, 63 varieties, 88–116
Vomiting of blood caused by Cobra-bites, 228

Worm snake, vermic. amphis., 98

Yellow fever, relation of to serpent-poison, 129, 131.

www.ingramcontent.com/pod-product-compliance
Lightning Source LLC
Chambersburg PA
CBHW031752230426
43669CB00007B/586